D0907171

Amino Acids in Psychiatric Disease

Series

David Spiegel, M.D.
Series Editor

Amino Acids in Psychiatric Disease

Edited by
Mary Ann Richardson, Ph.D.

Washington, DC
London

Note: The authors have worked to ensure that all information in this book concerning drug dosages, schedules, and routes of administration is accurate as of the time of publication and consistent with standards set by the U.S. Food and Drug Administration and the general medical community. As medical research and practice advance, however, therapeutic standards may change. For this reason and because human and mechanical errors sometimes occur, we recommend that readers follow the advice of a physician who is directly involved in their care or the care of a member of their family.

Books published by the American Psychiatric Press, Inc., represent the views and opinions of the individual authors and do not necessarily represent the policies and opinions of the Press or the American Psychiatric Association.

Copyright © 1990 American Psychiatric Press, Inc.
ALL RIGHTS RESERVED
Manufactured in the United States of America
First Edition 93 92 91 90 4 3 2 1

American Psychiatric Press, Inc.
1400 K St., N.W., Suite 1101, Washington, D.C. 20005

The paper used in this publication meets the minimum requirements of the American National Standard for Information Sciences—Permanence of Paper for Printed Library Materials, ANSI Z39.48–1984.   ∞

Library of Congress Cataloging-in-Publication Data

Amino acids in psychiatric disease/edited by Mary Ann Richardson.
    p.   cm. — (Progress in psychiatry series)
    ISBN 0-88048-186-2 (alk. paper)
    1. Mental illness—Physiological aspects. 2. Amino acids.
3. Amino acids—Therapeutic use. I. Richardson, Mary Ann. II. Series.
    [DNLM: 1. Amino acids—physiology. 2. Mental Disorders—physiopathology. WM 100 A517]
RC455.4.B5A46   1990
616.89'07—dc20
DNLM/DLC
for Library of Congress   89-18455
                                                              CIP

British Library Cataloguing in Publication Data

A CIP record is available from the British Library.

CONTENTS

Contributors

Ian N. Acworth, Ph.D.
Department of Brain and Cognitive Science, Massachusetts Institute of Technology, Cambridge, Massachusetts

Lars Bjerkenstedt, M.D.
Department of Psychiatry, Danderyds Hospital, Danderyd, Sweden

William A. Boggiano, M.D.
Walter Reed Army Medical Center; Uniformed Services University of Health Sciences, Washington, D.C.

Gerald Curzon, Ph.D., D.Sc.
Department of Neurochemistry, Institute of Neurology, London, England

Howard Kushner, Ph.D.
Nathan Kline Institute for Psychiatric Research, Orangeburg, New York

Timothy J. Maher, Ph.D.
Department of Pharmacology, Massachusetts College of Pharmacy, Boston, Massachusetts

Svend E. Møller, Ph.D.
Clinical Research Laboratory, St. Hans Hospital, Roskilde, Denmark

Michael A. Reveley, M.D., Ph.D., F.R.C.P.
Academic Department of Psychiatry, The London Hospital Medical College, University of London, London, England

Mary Ann Richardson, Ph.D.
Nathan Kline Institute for Psychiatric Research, Orangeburg, New York; New York University Medical Center, New York, New York

Raymond Suckow, Ph.D.
Analytical Psychopharmacology Division, New York State Psychiatric Institute; College of Physicians and Surgeons, Columbia University, New York, New York

Istvan Sziraki, Ph.D.
Institute for Drug Research, Budapest, Hungary

C. J. A. Taylor, M.B.B.S., B.Sc., M.R.C.P.
Academic Department of Psychiatry, The London Hospital Medical College, University of London, London, England

Herman M. van Praag, M.D., Ph.D.
Department of Psychiatry, Albert Einstein College of Medicine; Montefiore Medical Center, Bronx, New York

Rowland Whittaker, R.N.
Rockland Psychiatric Center, Orangeburg, New York

Richard J. Wurtman, M.D.
Department of Brain and Cognitive Science, Massachusetts Institute of Technology, Cambridge, Massachusetts

Simon N. Young, Ph.D.
Department of Psychiatry, McGill University, Montreal, Quebec, Canada

Introduction to the Progress in Psychiatry Series

The Progress in Psychiatry Series is designed to capture in print the excitement that comes from assembling a diverse group of experts from various locations to examine in detail the newest information about a developing aspect of psychiatry. This series emerged as a collaboration between the American Psychiatric Association's (APA) Scientific Program Committee and the American Psychiatric Press, Inc. Great interest is generated by a number of the symposia presented each year at the APA annual meeting, and we realized that much of the information presented there, carefully assembled by people who are deeply immersed in a given area, would unfortunately not appear together in print. The symposia sessions at the annual meetings provide an unusual opportunity for experts who otherwise might not meet on the same platform to share their diverse viewpoints for a period of 3 hours. Some new themes are repeatedly reinforced and gain credence, while in other instances disagreements emerge, enabling the audience and now the reader to reach informed decisions about new directions in the field. The Progress in Psychiatry Series allows us to publish and capture some of the best of the symposia and thus provide an in-depth treatment of specific areas that might not otherwise be presented in broader review formats.

Psychiatry is by nature an interface discipline, combining the study of mind and brain, of individual and social environments, of the humane and the scientific. Therefore, progress in the field is rarely linear—it often comes from unexpected sources. Further, new developments emerge from an array of viewpoints that do not necessarily provide immediate agreement but rather expert examination of the issues. We intend to present innovative ideas and data that will enable you, the reader, to participate in this process.

We believe the Progress in Psychiatry Series will provide you with an opportunity to review timely new information in specific fields of interest as they are developing. We hope you find that the excitement of the presentations is captured in the written word and that this book proves to be informative and enjoyable reading.

David Spiegel, M.D.
Series Editor
Progress in Psychiatry Series

Progress in Psychiatry Series Titles

The Borderline: Current Empirical Research (#1)
Edited by Thomas H. McGlashan, M.D.

Premenstrual Syndrome: Current Findings and Future
Directions (#2)
Edited by Howard J. Osofsky, M.D., Ph.D., and Susan J.
Blumenthal, M.D.

Treatment of Affective Disorders in the Elderly (#3)
Edited by Charles A. Shamoian, M.D.

Post-Traumatic Stress Disorder in Children (#4)
Edited by Spencer Eth, M.D., and Robert S. Pynoos, M.D., M.P.H.

The Psychiatric Implications of Menstruation (#5)
Edited by Judith H. Gold, M.D., F.R.C.P.(C)

Can Schizophrenia Be Localized in the Brain? (#6)
Edited by Nancy C. Andreasen, M.D., Ph.D.

Medical Mimics of Psychiatric Disorders (#7)
Edited by Irl Extein, M.D., and Mark S. Gold, M.D.

Biopsychosocial Aspects of Bereavement (#8)
Edited by Sidney Zisook, M.D.

Psychiatric Pharmacosciences of Children and Adolescents (#9)
Edited by Charles Popper, M.D.

Psychobiology of Bulimia (#10)
Edited by James I. Hudson, M.D., and Harrison G. Pope, Jr., M.D.

Cerebral Hemisphere Function in Depression (#11)
Edited by Marcel Kinsbourne, M.D.

Eating Behavior in Eating Disorders (#12)
Edited by B. Timothy Walsh, M.D.

Tardive Dyskinesia: Biological Mechanisms and Clinical
Aspects (#13)
Edited by Marion E. Wolf, M.D., and Aron D. Mosnaim, Ph.D.

Current Approaches to the Prediction of Violence (#14)
Edited by David A. Brizer, M.D., and Martha L. Crowner, M.D.

Treatment of Tricyclic-Resistant Depression (#15)
Edited by Irl L. Extein, M.D.

Depressive Disorders and Immunity (#16)
Edited by Andrew H. Miller, M.D.

Depression and Families: Impact and Treatment (#17)
Edited by Gabor I. Keitner, M.D.

Depression in Schizophrenia (#18)
Edited by Lynn E. DeLisi, M.D.

Biological Assessment and Treatment of Posttraumatic Stress Disorder (#19)
Edited by Earl L. Giller, Jr., M.D., Ph.D.

Personality Disorders: New Perspectives on Diagnostic Validity (#20)
Edited by John M. Oldham, M.D.

Serotonin in Major Psychiatric Disorders (#21)
Edited by Emil F. Coccaro, M.D., and Dennis L. Murphy, M.D.

Amino Acids in Psychiatric Disease (#22)
Edited by Mary Ann Richardson, Ph.D.

Family Environment and Borderline Personality Disorder (#23)
Edited by Paul Skevington Links, M.D.

Biological Rhythms, Mood Disorders, Light Therapy, and the Pineal Gland (#24)
Edited by Mohammad Shafii, M.D., and Sharon Lee Shafii, R.N., B.S.N.

Treatment Strategies for Refractory Depression (#25)
Edited by Steven P. Roose, M.D., and Alexander H. Glassman, M.D.

Combined Pharmacotherapy and Psychotherapy for Depression (#26)
Edited by Donna Manning, M.D., and Allen J. Frances, M.D

The Neuroleptic Nonresponsive Patient: Characterization and Treatment (#27)
Edited by Burt Angrist, M.D., and S. Charles Schulz, M.D.

Negative Schizophrenic Symptoms: Pathophysiology and Clinical Implications (#28)
Edited by John F. Greden, M.D., and Rajiv Tandon, M.D.

Neuropeptides and Psychiatric Disorders (#29)
Edited by Charles B. Nemeroff, M.D., Ph.D.

Introduction

As the title of this book suggests, amino acids are tools in understanding and treating psychiatric disease. This utility is demonstrated by the authors of this volume, who in presenting basic animal work, epidemiological investigations, clinical studies, and treatment trials, make use of the large neutral amino acids, as follows:

- To understand catecholamine synthesis and release
- To differentiate patients
- To define the pathophysiology of psychiatric disorders
- As treatments for these psychiatric disorders
- As predictors for the treatment efficacy of other pharmacologic agents in psychiatric disease

The monoamine neurotransmitters dopamine, norepinephrine, epinephrine, and serotonin are synthesized within the human brain from their precursors, the aromatic large neutral amino acids tyrosine and tryptophan. Phenylalanine, also an aromatic large neutral amino acid, relates directly to neurotransmitter synthesis as an inhibitor of the monoamine-hydroxylase enzymes. The other large neutral amino acids (histidine, threonine, valine, isoleucine, and leucine) affect precursor availability by influencing the rates at which the aromatic amino acids pass from blood to brain. Transport across the blood-brain barrier is believed to be the rate-limiting step for the penetration of amino acids into brain cells. A very high-affinity neutral amino acid transport system exists at the blood-brain barrier, with a particular affinity for the large neutral amino acids, the highest affinity being for phenylalanine. That the human brain is selectively vulnerable to the changes in plasma amino acid concentration is demonstrated by hyperaminoacidurias. One of these, the hyperphenylalaninemia phenylketonuria, shows this acute selective sensitivity of the brain to an excess of plasma phenylalanine in the almost exclusive involvement of the brain in this disease. The main features of the disorder are mental retardation, seizures, spasticity, and EEG irregularities. Behaviorally, brain effects are also evident; for example, phenylketonuric individuals are described as anxious, restless, subject to night terrors, destructive, noisy, hyperactive, irritable, and manifesting uncontrollable temper tantrums. In phenylketonuria, an excess of phenylalanine affects brain function by saturating the brain cells and by decreasing brain levels of the monoamines through 1) competing with the monoamine precursors tyrosine and tryptophan for uptake into the brain, and 2) inhibiting the monoamine-synthesizing enzymes, such as dopa decarboxylase, tyrosine hydroxylase, and tryptophan hydroxylase.

In the first chapter Drs. Acworth and Wurtman set the stage for their contribution and this volume by reviewing the animal work that defines the path from plasma large neutral amino acids to 1) the large neutral amino acid ratio, 2) brain levels of the large neutrals, and 3) the synthesis of the monoamines from their precursor amino acids. These authors describe the animal work on tyrosine as a precursor of catecholamine metabolism. They point to the particular relevance to clinical populations of studies that demonstrate that while tyrosine administration generally produces a transient increase in the release of catecholamines, there are conditions under which this release can be increased and extended, for example, by 1) pretreating with haloperidol, reserpine, or yohimbine; 2) subjecting rats to cold stress; and 3) administering tyrosine to spontaneously hypertensive rats.

Other cited studies that employ investigative techniques of in vitro brain slices and in vivo microdialysis suggest further that tyrosine may not be concentrated within dopaminergic neurons but rather is mobilized at need; in periods of extreme neuronal activity, catecholamine synthesis may be inhibited by the availability of tyrosine.

In the second chapter, Dr. Curzon focuses on the role of special situational variables and individual vulnerability in affecting brain amino acid concentrations, primarily for tryptophan. The work presented demonstrates that tryptophan transport to the brain is increased by food deprivation, exercise, immobilization, and a large carbohydrate meal. The mechanisms by which tryptophan transport is increased are 1) for food deprivation and for exercise, by increasing nonprotein-based plasma tryptophan; 2) for immobilization by altering large neutral amino acid kinetics; and 3) for a large carbohydrate meal, by decreasing plasma concentrations of the other large neutral amino acids that compete with tryptophan for uptake into the brain. Dr. Curzon cautions against relying solely on the large neutral amino acid ratio as an index of brain levels of all amino acids and against extending the findings in the studies he reports and in others beyond the special populations studied.

Dr. Young, in his chapter, concentrates on the role of tryptophan in the treatment of affective disorders, pathological aggression, insomnia, and pain. He discusses a physiological limitation on serotonin precursor loading based on the facts that 1) tryptophan hydroxylase is normally half saturated, and thus tryptophan loading can do no more than double the rate of serotonin synthesis; and 2) increased synthesis does not necessarily mean an increase of serotonin released from neurons. An approach that has been used to bypass this limitation and effect a large release of serotonin independent of neuronal firing is the joint administration of a monoamine oxidase inhibitor (MAOI) and tryptophan. The MAOI then acts to inhibit the degradation of serotonin. Dr. Young next reviews the literature on tryptophan efficacy in depression. He believes the work he presents demonstrates that tryptophan is a useful antidepressant in either mildly or moderately depressed patients.

Dr. van Praag provides an incisive approach to the underlying structures of the psychological and biological dysfunctions of psychiatric disease and to demonstrating the malleability of these structures to precursor monoamine manipulation. In his chapter he outlines a multiaminergic hypothesis of depression, the two components of which are 1) a deficiency in motor functioning and level of initiative with a core biological dysfunction of diminished dopamine metabolism; and 2) anhedonia, an inability to couple a memory or an anticipation of reward to an action, the biological dysfunction for this symptom being deficient noradrenergic activity. Based on this theoretical framework and on previous clinical studies, Dr. van Praag presents a multipharmaceutical approach that regulates monoamine precursors to treat his outlined psychological and biological dysfunctions of depression.

Dr. Møller presents a unique approach to the treatment of psychiatric disease. By using plasma amino acids as predictors of treatment efficacy in depression, he provides a useful clinical application of the following ideas:

1. Serotonin brain levels are a function of brain tryptophan.
2. Brain noradrenaline derives in part from brain tyrosine concentration.
3. Brain tryptophan and tyrosine levels are determined by uptake from the L-carrier transport system across the blood-brain barrier.
4. The competition among the large neutral amino acids determines each one's uptake into the brain.
5. This competition is best represented by the ratio of the plasma level of each large neutral amino acid to the sum of the plasma levels of the others.

The work presented in this chapter, which is compiled from several studies, demonstrates that the tryptophan and tyrosine pretreatment ratios were predictors of therapeutic response to antidepressant treatments in 150 psychiatric inpatients suffering from a major depressive episode. These ratios proved to be significantly related to efficacy, whereas diagnostic variables and serum levels of antidepressants were not.

In Chapter 6 Dr. Maher alters direction from the previous chapters to focus on the essential aromatic amino acid phenylalanine. He outlines the dual dose-dependent role of phenylalanine as a catecholamine precursor and as an inhibitor of catecholamine synthesis and/or release. Dr. Maher emphasizes the importance of the large neutral amino acid ratio in understanding the effect of phenylalanine on brain function. He cautions readers on the differences between rodent and human amino acid metabolism and the risk inherent in drawing inferences to humans from rodent models. Dr. Maher suggests that future work on the neurological, behavioral, or cognitive effects of increases in plasma phenylalanine be designed to capture subtle chronic effects rather than the gross measures sometimes seen in this work.

The next chapter defines my investigation into phenylalanine metabolism as a risk factor for the development of tardive dyskinesia in schizophrenic patients. The reported study was stimulated by previous work indicating that phenylketonuric patients treated with neuroleptics were at particular risk for the development of tardive dyskinesia. The positive findings of the work on phenylalanine and tardive dyskinesia were (a) that a postprotein-loading phenylalanine level was significantly associated with tardive dyskinesia status and severity and (b) that the phenylalanine/large neutral amino acid ratio was a risk factor for tardive dyskinesia development and also was associated significantly to its severity. These findings provide new insights into the pathophysiology of the disorder and supply data for developing a treatment for tardive dyskinesia based on regulating plasma amino acid.

Dr. Bjerkenstedt continues the focus on schizophrenic patients and reports on the work of two investigations studying amino acid differences between patients and normal controls. The first of these studies collected plasma amino acids and cerebrospinal fluid (CSF) homovanillic acid. The second study, completed by a group of investigators including Dr. Bjerkenstedt (with Dr. L. Hagenfeldt as first author), examined amino acid transport in fibroblasts in schizophrenic patients and in controls. Dr. Bjerkenstedt sees both of these studies as implicating disturbances in tyrosine transport in the pathophysiology of schizophrenia.

In the final chapter Drs. Taylor and Reveley report further on schizophrenic patients, beginning with a review of the studies that postulate a role for amino acids in the etiology and pathophysiology of schizophrenia. The amino acids covered include phenylalanine, gamma-aminobutyric acid (GABA), and glutamic acid. The route of amino acid transport mechanisms in this postulated association between amino acid function and schizophrenia is discussed. The authors also review their own work examining the relationship of CSF concentrations of amino acids to ventricular enlargement in schizophrenia.

It is hoped that this volume, by broadening the audience for the studies presented, will both generate more research on amino acids and provide information to clinicians that they will value. The authors of this volume believe that further research in amino acids and psychiatric disease may foster a better understanding of these disorders and may provide a data base for the development of more effective, less troublesome treatment modalities based on plasma amino acid regulation.

I would like to acknowledge the contributions of all the authors to this book and the assistance of Frances Simpson in its preparation.

This book is dedicated to the memory of my mother, Anna Maria Sanchirico (1898–1988), and my brother, Anthony Joseph Sanchirico (1923–1988).

Mary Ann Richardson, Ph.D.

Chapter 1

Precursor Control of Catecholamine Metabolism

Ian N. Acworth, Ph.D.
Richard J. Wurtman, M.D.

Chapter 1

Precursor Control of Catecholamine Metabolism

The mammalian brain uses several diverse classes of compounds as neurotransmitter substances, including, among others, amino acids, monoamines, and peptides. The catecholamines dopamine, norepinephrine, and epinephrine and the indole-alkylamine serotonin (5-hydroxytryptamine [5-HT]) constitute the monoamines, a group of neurotransmitters that are synthesized within the brain from their precursor amino acids, tyrosine and tryptophan, respectively. These amino acids, in turn, must be derived from the circulation.

PLASMA AMINO ACIDS AND THE LARGE NEUTRAL AMINO ACID RATIO

The concentrations of tyrosine and tryptophan, as well as of the other large neutral amino acids (LNAAs) in plasma, are subject to wide variations (Fernstrom et al. 1979; Maher et al. 1984). They change when people or animals receive various drugs such as antidepressants (Badawy and Evans 1981), psychoactive compounds (Valzelli et al. 1980), or the amino acids themselves (Fernstrom and Wurtman 1971a; Glaeser et al. 1979) and in disease states affecting amino acid metabolism (e.g., phenylketonuria, maple syrup urine disease, hepatic encephalopathy) (Curzon 1980; Curzon et al. 1982; Stanbury et al. 1966). A change of major physiological relevance occurs in association with eating (Fernstrom et al. 1979; Maher et al. 1984) or with strenuous exercise (Acworth et al. 1986; Conlay et al. 1989; Decombaz et al. 1979).

An understanding of the ability of plasma amino acid patterns to influence brain monoamine metabolism comes originally from studies

These studies were supported in part by grants from the United States Air Force, the National Aeronautics and Space Administration, and the Center for Brain Sciences and Metabolism Charitable Trust.

of macronutrient effects on 5-HT metabolism. The ability of nutrients to affect brain composition was first demonstrated in 1971, in experiments in which rats consuming meals containing carbohydrate and fat (i.e., lacking protein) were found soon thereafter to have increased brain levels of the essential and scarce amino acid tryptophan (Fernstrom and Wurtman 1971b). The rats also exhibited increased brain levels of 5-HT and its metabolite 5-hydroxyindoleacetic acid (5-HIAA) (Fernstrom and Wurtman 1971b)—findings compatible with the view that the rise in brain tryptophan levels had increased the substrate saturation of tryptophan hydroxylase, the initial and rate-limiting enzyme in 5-HT synthesis (Bradford 1986). The change in levels of 5-HIAA was taken as indirect evidence that the release of 5-HT had increased in parallel (see below). The increase in brain tryptophan levels was initially thought to have resulted from the small increase in plasma tryptophan concentration that occurs in rats but not in humans after carbohydrate intake. It was unclear at that time why both plasma and brain tryptophan levels should *rise* following ingestion of a meal *deficient* in tryptophan, especially as insulin release (in response to carbohydrate) was know to *lower* the plasma levels of the other LNAAs (Fernstrom and Wurtman 1972b). This unusual response to carbohydrate intake became understandable several years later when the binding of 5-HT to serum albumin was taken into account (Knott and Curzon 1972; Madras et al. 1974; McMenamy and Oncley 1958). Insulin, the major antilipolytic hormone, increases the transfer of free fatty acids from serum albumin to adipocytes. This transfer, in turn, diminishes the free fatty acid content of the albumin, *increasing* its affinity for tryptophan (Madras et al. 1974). Plasma levels of free (i.e., nonalbumin-bound) tryptophan actually fall in response to insulin, but this decrease is compensated by the rise in albumin-bound tryptophan—a moiety that is almost as accessible to brain as the "free" fraction (Yuwiler et al. 1977).

When rats were fed protein-rich meals, there appeared at first to be no relationship between plasma and brain tryptophan levels; although plasma tryptophan levels rose (derived from some of the tryptophan in the meal's protein), brain tryptophan and 5-HT levels either failed to rise or, if the meal contained sufficient protein, actually fell (Fernstrom and Wurtman 1972a). It was suggested that this phenomenon reflected competition for brain uptake between tryptophan and the other LNAAs that are more abundant in protein, an explanation supported by Pardridge and Oldendorf's (1986) subsequent characterization of the mechanism that transports the LNAAs across the blood-brain barrier (Pardridge 1986). Tryptophan shares the LNAA carrier—a facilitated diffusion system—with other LNAAs (tyrosine, phenylalanine, histidine, leucine, isoleucine, valine, threonine, methionine); the carrier's affinity for tryptophan (0.052 mM) is less than that of phenylalanine (0.032 mM) but in the same order as those of tyrosine, leucine, and methionine (Pardridge 1986). The carrier's net affinity for tryptophan, the amino acid's net flux into the brain, and, after sufficient time, even the changes in brain tryptophan levels that follow a meal, can all be estimated by calculating a "plasma tryptophan ratio": the ratio of the plasma tryptophan concentration to the summed concentrations of the other LNAAs that bind to the carrier with a reasonably high affinity (see above) (Fernstrom and Wurtman 1972a). Because all dietary proteins are considerably richer in the competing LNAAs than in tryptophan, which generally constitutes only 1.0–1.5% of protein, consumption of a meal that is rich in protein (e.g., 30–40% of calories) can cause the plasma tryptophan *ratio* to fall, even as plasma tryptophan *levels* rise. Similar competitive mechanisms mediate the fluxes of tryptophan and other LNAAs between the brain's extracellular space and individual neurons; however, this competition probably is not limiting, because the V_{max} of transport at this locus is tenfold that at the blood-brain barrier (Pardridge 1986). Moreover, similar plasma ratios predict brain levels of tyrosine and of each of the other LNAAs after treatments that modify plasma amino acid patterns (Fernstrom and Faller 1978).

At that time it was also shown that consumption of supplemental tyrosine could affect the metabolism of the catecholamines (Wurtman et al. 1974). These responses, however, were less well characterized than the effects of supplemental tryptophan on 5-HT. Evidence to date suggests that, apart from the monoamines, only a few other neurotransmitters are subject to precursor control: acetylcholine (Cohen and Wurtman 1975; Haubrich et al. 1975), histamine (Schwarz et al. 1972), and glycine (Maher and Wurtman 1980). Pharmacological doses of histidine or threonine can elevate brain concentrations of their respective neurotransmitter products, histamine (Schwarz et al. 1972) and glycine (Maher and Wurtman 1980), and choline availability is a major factor in determining how much acetylcholine is provided and released when particular cholinergic neurons fire (Maire and Wurtman 1984).

It cannot be stated for certain whether the excitatory amino acid neurotransmitters (i.e., aspartate, glutamate) are under precursor control, because the identities of the precursors for these compounds within *glutamatergic* or *aspartatergic* neurons await discovery. Similarly, experiments have not yet been done to assess the effects on GABA synthesis caused by raising brain levels of GABA's precursor, glutamate; little exogenous glutamate crosses the blood-brain barrier, and the amino acid thus fails to elevate its own brain levels substantially (Liebschutz et al. 1977).

CONDITIONS FOR PRECURSOR CONTROL OF NEUROTRANSMITTER METABOLISM

In order for a precursor-mediated increase in neurotransmitter synthesis to affect the neurotransmitter's release, the neuron involved must continue to fire at a reasonably high frequency. Feedback processes operating via both short presynaptic (autoreceptor) or long postsynaptic loops may prevent this accelerated firing from happening. The absence of overt physiologic effects after healthy humans consume tyrosine (Glaeser et al. 1979) or choline (as phosphatidylcholine or "lecithin") (Wood and Allison 1981) suggests that these controlling mechanisms are operating. (Serotonergic neurons seem to have greater freedom to respond to supplemental tryptophan, as assessed neurochemically and by the physiological and behavioral effects of this amino acid.) However, circumstances do exist in which receptor-mediated feedback mechanisms apparently are not activated by increases in catecholamine or acetylcholine release, thus allowing precursor administration to amplify neurotransmission. Such circumstances include, among others, *neurodegenerative disorders*, in which surviving neurons will continue to exhibit accelerated firing frequencies so long as the total quantity of neurotransmitter released from the afflicted tract or nerve is less than adequate; *physiological states* in which a prolonged increase in neurotransmitter is required (e.g., to sustain blood pressure in hypovolemia or blood glucose after insulin administration); drug treatments that cause neurotransmitter "release" to be inadequate (e.g., release of tyrosine after reserpine treatment [Sved et al. 1979a]); and within *neurons not "hard-wired" into multisynaptic feedback loops* or within those serving as components of *positive* feedback loops.

TYROSINE EFFECTS ON CATECHOLAMINE SYNTHESIS AND RELEASE

Early Studies

Because tyrosine administration generally does not increase brain dopamine or norepinephrine levels among otherwise untreated animals, most investigators assumed that catecholamine synthesis was not under precursor control (Table 1-1). This apparent lack of precursor responsiveness was perhaps surprising, in view of the fact that plasma tyrosine levels were known to greatly increase after protein intake (Fernstrom et al. 1979; Maher et al. 1984) or tyrosine administration (Glaeser et al. 1979); that the LNAA transport system was as capable of ferrying tyrosine as tryptophan across the bloodbrain barrier (Pardridge 1977); and that tyrosine hydroxylase, which catalyzes the rate-limiting step in catecholamine synthesis, was, like tryptophan hydroxylase, not likely to be saturated with its amino acid substrate in vivo (Sved 1983). The possibility was considered that a small, rapidly turning-over catecholamine pool that was responsive to tyrosine did exist, but that this pool was too small, in relation to total catecholamine stores, to allow detection of an increase in its size after tyrosine administration. Initial studies were therefore conducted to determine whether catecholamine *synthesis* or *release*, assessed independently of brain catecholamine levels, might be affected by changes in brain tyrosine concentrations.

At first, catecholamine synthesis was estimated by following the rate at which L-dopa accumulated in brains of animals pretreated acutely with an inhibitor of aromatic amino acid decarboxylase (Badawy and Williams 1982; Carlsson and Lindqvist 1978; Wurtman et al. 1974). Tyrosine administration did sometimes accelerate L-dopa's accumulation. For example, Wurtman et al. (1974) reported that a twofold increase in brain tyrosine levels could increase L-dopa levels by 13% in the whole brain. However, using a similar system, Carlsson and Lindqvist (1978) reported no effect of tyrosine administration on L-dopa accumulation in the whole brain. If individual brain regions were examined, tyrosine had little effect on L-dopa accumulation in the striatum or limbic forebrain; but given at a dose that raised tissue tyrosine levels tenfold, tyrosine did increase cerebral L-dopa accumulation 25%. The rate of catecholamine synthesis could more readily be diminished by administration of the other LNAAs (e.g., tryptophan, valine, or parachlorophenylalanine) that compete with tyrosine for transport across the blood-brain barrier, and whose administration thus lowers brain tyrosine levels (Wurtman et al. 1974). These experiments allowed an estimation of the K_m of tyrosine hydroxylation in vivo and showed that this process could indeed be affected when tyrosine levels varied within their normal physiological range. (Extrapolation of these findings to the physiological state was complicated because the inhibitor not only diminishes catecholamine synthesis but also suppresses both the end-product inhibition of tyrosine hydroxylase and the release of newly formed catecholamine into the synapse.)

A technique allowing estimation of catecholamine synthesis in the presence of varying concentrations of brain tyrosine without disturbing that synthesis was then developed. Investigators used the brain

levels of the major metabolites of dopamine (3,4-dihydroxy-
phenylacetic acid [DOPAC]; 4-hydroxy-3-methoxyphenylacetic acid
[homovanillic acid, HVA]) or norepinephrine (3-methoxy-4-hydrox-
yphenylglycol sulfate [MHPG-SO4]) in animals given tyrosine as an
index of catecholamine synthesis (when catecholamine levels
remained unchanged). Interestingly, the administration of even large
doses of tyrosine as either the pure amino acid or the more soluble
methyl ester to otherwise untreated animals had no effects on the
levels of these metabolites in most instances (Melamed et al. 1980;
Scally et al. 1977; Sved and Fernstrom 1981; Sved et al. 1979b).
Gibson and Wurtman (1978), however, did find that systemically
administered tyrosine resulted in a 15% increase in whole brain
MHPG-SO4 levels in rats pretreated with probenecid (to block
metabolite egress from the brain. Again, systemic administration of
competing LNAAs reduced both tissue tyrosine levels and the rate of
catecholamine synthesis.

However, if the experimental animals were given an additional
treatment designed to accelerate the firing of dopaminergic or
noradrenergic tracts, then the supplemental tyrosine markedly enhanced the accumulation of the catecholamine metabolites (Table
1-1). For example, when rats were given the dopamine antagonist
haloperidol, which accelerates the firing frequency of the nigrostriatal
pathway (Bunney et al. 1973) and lowers striatal tyrosine levels
(Westerink and Wirix 1983), then systemic administration of tyrosine
increased catecholamine synthesis; tissue levels of HVA varied directly
with those of striatal tyrosine, whereas dopamine levels remained
constant (Scally et al. 1977). These initial observations formed the
basis for the hypothesis that catecholaminergic neurons become
tyrosine-sensitive when they are physiologically active, and lose this
capacity when quiescent.

The ability of additional tyrosine to enhance catecholamine synthesis in, and release from, only those neurons that are rapidly firing
has been shown using a variety of pharmacological as well as
physiological treatments. Tyrosine administration increases brain
levels of dopamine metabolites when animals are pretreated with
reserpine (Sved et al. 1979a), spiperone (Fuller and Snoddy 1982),
or amfonelic acid (Fuller and Snoddy 1982), which are thought to
increase the firing frequency of nigrostriatal neurons. Also, supplemental tyrosine can enhance ipsilateral dopamine metabolism in
rats with unilateral nigrostriatal lesions. Melamed et al. (1980)
reported that following a 6-hydroxydopamine-induced nigrostriatal
lesion (which destroyed more than 80% of that tract, and thus
accelerated the firing of the surviving neurons [Agid et al. 1973]),

tyrosine administration increased dopamine release on the lesioned side (as estimated by the ratios of DOPAC or HVA to dopamine, or to tyrosine hydroxylase activity), but not on the intact side (using the same indices). This effect could be blocked by coadministration of valine (another LNAA that effectively competes with tyrosine for transport into the brain) and was not associated with changes in tissue dopamine levels.

Systemically administered tyrosine need not affect the various catecholaminergic pathways in an identical fashion. For example, Sved (1980) found that tyrosine administration normally does not alter the rate of dopamine synthesis (L-dopa accumulation after decarboxylase inhibition) in the striatum or in the median eminence. However, following intracerebral injection of ovine prolactin, a treatment that specifically activates tuberoinfundibular neurons projecting to the median eminence (Annunziato 1979; Annunziato and Moore 1978; Moore et al. 1978), tyrosine administration increased the rate of dopamine synthesis in the median eminence, but not in the striatum.

Tyrosine supplementation also increases the levels of dopamine metabolites in light-activated retinas in vivo but not when the animals are in darkness (Gibson et al. 1983). Retinal dopamine synthesis and turnover can also be altered when rats consume protein or tyrosine-supplemented diets (Gibson 1986, 1988). Low doses of tyrosine also increase dopamine synthesis in brain neurons that exhibit high firing frequencies and bursting activity, such as the mesoprefrontal dopaminergic neurons (Tam and Roth 1984). Interestingly, tyrosine administration can also increase dopamine synthesis (Sved and Fernstrom 1981) in neurons that are not firing frequently, but in which tyrosine hydroxylase has been activated pharmacologically (e.g., in nigrostriatal neurons of rats pretreated with γ-butyrolactone, which inhibits their firing but activates tyrosine hydroxylase by decreasing presynaptic inhibition [Morgenroth et al. 1976; Walters and Roth 1974, 1976; Walters et al. 1972, 1973]).

Tyrosine administration can also affect norepinephrine metabolism. For example, the level of the chief metabolite of norepinephrine, MHPG-SO₄, is elevated in the whole brain when cold-stressed animals are administered tyrosine (Gibson and Wurtman 1978). Also, tyrosine supplementation increases the level of MHPG-SO4 in the brain stem of spontaneously hypertensive rats, but not in normotensive controls or in spontaneously hypertensive rats also given valine (Sved et al. 1979b; Yamori et al. 1980). (It is interesting that tyrosine administration apparently fails to affect dopamine metabolism in any brain area of the spontaneously hypertensive rat.) Increases in MHPG-SO4 after tyrosine administration are seen in animals

Table 1-1. Tyrosine administration and catecholamine synthesis and release

Tissue	Treatment	Biochemical index	Tyrosine effect (%)	Reference(s)
Whole Brain	Tyrosine alone	L-Dopa	+13	Wurtman et al. 1974
Whole Brain	Tyrosine alone	L-Dopa	+31	Badawy and Williams 1982
Cerebellum	Tyrosine alone	L-Dopa	+25	Carlsson and Lindqvist 1978
Whole Brain	Tyrosine/probenecid	MHPG-SO$_4$	+15	Gibson and Wurtman 1978
Striatum[a]	Haloperidol	DOPAC, HVA	+60	Scally et al. 1977
Striatum/limbic forebrain[a]	Haloperidol	L-Dopa	+15	Carlsson and Lindqvist 1978; Westerink and Wirix 1983
Striatum[a]	6-Hydroxydopamine-induced lesion	DOPAC, HVA	+60	Melamed et al. 1980

Striatum[a]	γ-Butyrolactone	L-Dopa	+25	Sved and Fernstrom 1981
Hypothalamus[a]	Reserpine	DOPAC, HVA	+40	Sved et al. 1979a
Median eminence[a]	Prolactin	L-Dopa	+30	Sved 1980
Whole Brain	Amfonelic acid, spiperone	DOPAC	+30	Fuller and Snoddy 1982
Whole Brain[a]	Cold stress	MHPG-SO$_4$	+70	Gibson and Wurtman 1978
Whole Brain[a]	SHR	MHPG-SO$_4$	+40	Sved et al. 1979b
Forebrain/brain stem	SHR	MHPG-SO$_4$	+15	Yamori et al. 1980
Whole Brain	Yohimbine	MHPG-SO$_4$	+35	Gibson 1977
Hypothalamus/hippocampus	Tail shock	MHPG-SO$_4$	+40	Reinstein et al. 1984

Note. MHPG-SO$_4$ = 3-methoxy-4-hydroxyphenylglycol sulfate; DOPAC = 3,4-dihydroxyphenylacetic acid; HVA = homovanillic acid; SHR = spontaneously hypertensive rats.
[a]Limited effect on biochemical index when tyrosine administered alone.

pretreated with the α_2 antagonist yohimbine (Gibson 1977) or in those animals subjected to tail-shock stress (Reinstein et al. 1984).

The use of brain levels of catecholamine metabolites as indices of synthesis and release of these metabolites involves assumptions that may or may not be valid (e.g., that the flux of these metabolites out of the brain is unaffected by the experimental treatment). Wisdom requires that before the hypothesis that tyrosine levels affect catecholamine synthesis can be accepted, it should be supported by direct evidence on the release of the transmitter.

Studies Using Brain Slices

Much evidence has now been obtained using both in vitro brain slice studies and in vivo microdialysis studies (see next section). Milner and Wurtman (1985, 1986) examined the effects of adding tyrosine to the superfusate on dopamine release from electrically stimulated rat striatal slices (which were also exposed to the dopamine reuptake blocker nomifensine). Slices were subjected to electrical pulses (20 Hz, 2 milliseconds) of various train lengths (either 600 pulses [30 seconds] or 1,800 pulses [90 seconds]), and the amount of dopamine released into the superfusate medium was correlated with tyrosine concentration. When tissue was superfused in Krebs bicarbonate buffer lacking tyrosine and was subjected to two trains of pulses (1,800 pulses) delivered 60 minutes apart, the amount of dopamine released during the second stimulation period decreased by 20 to 25% (Figure 1-1). However, if tyrosine was included in the medium (50 μM), then dopamine release remained unaltered after both stimulation periods. Milner and Wurtman (1985) then showed that a tyrosine concentration in the superfusate of at least 20 μM was needed to maintain dopamine release during both stimulation periods in slices stimulated for 30 seconds, while at least 40 μM was needed in tissue stimulated for 90 seconds (Figure 1-2). Tissue that had been stimulated in tyrosine-free medium also showed a marked decrease in tyrosine content (up to 50%), as well as in tissue dopamine levels (25%). (The ability of supplemental tyrosine to maintain catecholamine release in the face of repeated firing apparently is not characteristic of all catecholaminergic neurons; norepinephrine release from superfused hypothalamic slices was not sustained by adding tyrosine to the medium [Irie and Wurtman 1987].) Because dopaminergic terminals constitute only a small proportion of the total cellular mass of the striatum, this decrease in tissue tyrosine poststimulation either represented a depletion in noncatecholaminergic cells, as well as dopaminergic cell terminals, or showed that most of the tyrosine in the striatum is located in dopaminergic neurons.

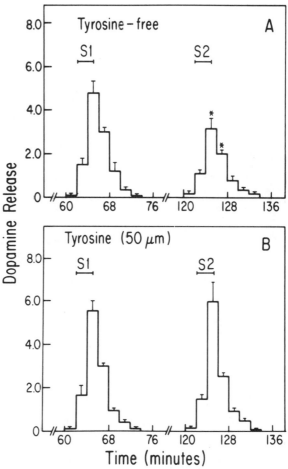

Figure 1-1. Release of endogenous dopamine evoked by electrical stimulation from rat striatal slices, expressed as percentage released of final tissue content. (A) Tyrosine-free medium; (B) tyrosine-supplemented medium (50 μM). S1 and S2 were identical trains of 1,800 pulses (60 mA, 2 milliseconds, 20 Hz) delivered 60 minutes apart. Superfusate was collected every 2 minutes and assayed for dopamine by alumina extraction and high-performance liquid chromatography with electrochemical detection. Data were analyzed by paired Student's *t* tests, and values are shown as mean ± SEM for four experiments. Asterisk denotes $P < .05$ when compared to equivalent fractions from S1 (Milner and Wurtman 1986).

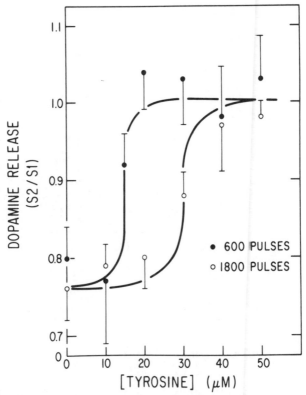

Figure 1-2. Effect of tyrosine on stimulus-evoked release of endogenous dopamine from superfused rat striatal slices. Trains of electrical pulses (60 mA, 2 milliseconds, 20 Hz) were delivered for 30 (600 pulses) or 90 (1,800 pulses) seconds. Dopamine release during S1 averaged 3.92 ± 0.26 pmoles/mg (600 pulses) and 7.91 ± 0.59 pmoles/mg (1,800 pulses), thus declining in the amount released per pulse by approximately 33%. Total dopamine released during the first train (S1) was compared with that released by the second identical train applied 30 minutes later (S2/S1). Tyrosine dose-dependently increased the release of dopamine during S2, with approximate S2/S1 unity at tyrosine concentrations of 20 μM (600 pulses) or 40 μM (1,800 pulses). Comparison of S2/S1 ratio for the two trains by Student's t test showed significant differences ($P < .05$) in dopamine release evoked by S2 during superfusion with tyrosine concentrations of 20 and 30 μM. Each point represents the mean ± SEM of four to seven animals (Milner and Wurtman 1985).

Using unilateral 6-hydroxydopamine substantia nigra lesions, Milner et al. (1987) reported that although dopamine levels in the ipsilateral striatum were decreased by more than 95%, tyrosine levels in that area remained unchanged; this finding suggests that tyrosine is not particularly concentrated within dopaminergic neurons and that they may mobilize it from nondopaminergic neurons if the need arises. Physiologically, it is unlikely that catecholaminergic neurons will experience periods of major tyrosine deprivation in vivo, because the brain is continuously furnished with a constant supply of tyrosine in the plasma. However, the rate at which tyrosine diffuses from the plasma into the neuron is slowed by its poor water solubility and its need to be carried across the blood-brain barrier (Oldendorf 1971) and the neuronal cell membrane (Guroff et al. 1961). The possibility, therefore, remains that, under prolonged periods of extreme neuronal activity, catecholamine synthesis may be limited by tyrosine supply.

Milner et al. (1986) reported that catecholamine release from superfused striatal slices can also be influenced by phenylalanine. In the absence of tyrosine, phenylalanine (25 µM) partially sustained dopamine release, but less well than did an equimolar concentration of tyrosine. But in the presence of tyrosine (50 µM), high levels of phenylalanine (> 200 µM) *inhibited* dopamine release into the superfusate by 30 to 40%. This inhibition was not associated with changes in tissue levels of tyrosine or dopamine, nor was it mimicked by adding high concentrations of tyrosine or leucine to the medium. It probably reflected both decreased tyrosine transport across the blood-brain barrier (Pardridge 1986) and direct competitive inhibition by phenylalanine of tyrosine hydroxylase (Ikeda et al. 1967; Shiman et al. 1971).

Microdialysis Studies

Microdialysis is one of several in vivo techniques now being used to explore brain neurotransmitter metabolism in both anesthetized and awake animals. (Other methods include voltammetry, push-pull cannulation, and cortical-cup procedures [see Marsden (1984) for a review].) Briefly, dialysis has been likened to the placement of a synthetic capillary (in this case a semipermeable membrane at the end of a "probe") within a tissue. Perfusion of the probe with an artificial cerebrospinal fluid allows sampling of low-molecular-weight compounds from around the probe that enter it by diffusion (for review, see Ungerstedt 1984).

Recently, we have used microdialysis to explore the effects of various amino acids on the release and metabolism of dopamine in the striatum and nucleus accumbens of anesthetized animals.

Tyrosine, when administered intraperitoneally in doses of 50–200 mg/kg, transiently, but significantly, increased dopamine levels in striatal dialysates by 28 to 45% above basal values (Acworth et al. 1988) (Figure 1-3). The fact that tyrosine administration can increase dopamine release in otherwise untreated animals suggests that dialysis is able to measure a small pool of newly synthesized, preferentially released dopamine that is too small a proportion of the total dopamine pool to be measured using more conventional neurochemical

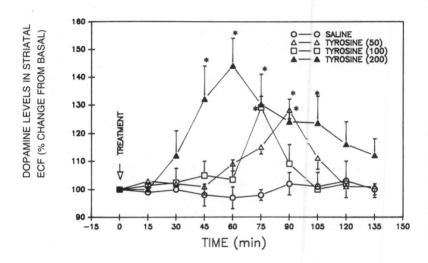

Figure 1-3.   Effects of intraperitoneal administration of saline or various doses of tyrosine (50, 100, or 200 mg/kg) on striatal extracellular fluid (ECF) levels of dopamine. Groups of rats (n = 5) were anesthetized with α-chloralose/urethane (0.05/0.5 g per kg ip). Microdialysis probes were placed acutely in the right striatum (A: +0.5; R: 2.5; V: −7) and were perfused with artificial CSF at 1.5 μl/minute. After injury release (100–120 minutes post probe implantation) and when basal levels of dopamine in three consecutive collections varied by less than 8%, rats then received either saline or tyrosine (50, 100, or 200 mg/kg, 4 ml/kg) as a suspension in saline. Samples were analyzed every 15 minutes for dopamine by high-performance liquid chromatography with electrochemical detection. Groups of five animals were used for each drug treatment. Vertical bars represent standard error of the mean. Statistical significance was measured using unpaired Student's t test. Asterisk denotes $P < .05$. (Acworth et al. 1988)

methods such as measuring dopamine levels in homogenates (see above). An equivalent dose of tyrosine (200 mg/kg) is more effective at increasing dopamine release in the nucleus accumbens than in the striatum (104% vs. 45%) (During et al. 1988a). This finding agrees with other neurochemical data showing that the nucleus accumbens exhibits a higher dopamine turnover (Beal and Martin 1985) and rate of synthesis (Anden et al. 1983). In each of the previous cases, the effect of supplemental tyrosine on dopamine release was short lived, suggesting that there are feedback mechanisms, perhaps based on autoreceptors or transsynaptic control of firing frequencies, that are activated to restore dopamine release to basal levels. In agreement with this explanation, the duration of tyrosine's (100 mg/kg) effect on dopamine release in the striatum was prolonged and the effect potentiated when animals were pretreated with the dopamine antagonist haloperidol (2 mg/kg) (During et al. 1988b). Also, if surviving nigrostriatal neurons were forced to fire more frequently— for example, after producing partial lesions with 6-hydroxydopamine—then systemically administered tyrosine (100 mg/kg) transiently increased dopamine levels in striatal ECF to 240% of basal values (During et al. 1988b) as compared with 30% in unoperated controls.

Phenylalanine can also affect striatal dopamine release as measured by microdialysis. Its effects vary with dose: a low dose (200 mg/kg) increases dopamine release by 59%; larger doses fail to affect dopamine release (500 mg/kg) or actually inhibit dopamine release by 26% (1,000 mg/kg) (During et al. 1988b) (Figure 1-4). A possible explanation of this triphasic effect of phenylalanine on striatal dopamine release is as follows: rodents have highly efficient hepatic conversion of phenylalanine to tyrosine (Moldawer et al. 1983), so that low doses of phenylalanine preferentially elevate plasma tyrosine levels. Tyrosine is, therefore, more effective at competing for access to the LNAA carrier (Pardridge 1977) for entry into the brain, where it can act as a substrate for dopamine synthesis. At somewhat higher doses, the hydroxylation of phenylalanine becomes less efficient, so that the plasma tyrosine/phenylalanine ratio approaches unity and neither amino acid has advantage at the LNAA carrier for transport into the brain. At the highest dose of phenylalanine used, the decrease in dopamine release is likely to reflect at least two processes, for example, the reduction in the plasma tyrosine/LNAA ratio and the consequent reduction in brain tyrosine levels (Fernstrom and Faller 1978) and the inhibition of tyrosine hydroxylase (Ikeda et al. 1967; Katz et al. 1976).

Finally, when valine (200 mg/kg) is given systemically to rats, striatal ECF levels of dopamine remain unaltered for at least 2 hours (During et al. 1989). This finding suggests that the effects of tyrosine and phenylalanine on dopamine release are not due to some diffuse stress response.

The average concentration of dopamine in striatal ECF is approximately 4–5 nM (Church et al. 1987; During et al. 1988a; Imperato and Di Chiara 1984; Zetterstrom et al. 1986), or less than one five-hundredth of those of its metabolites DOPAC and HVA. This finding suggests that the levels of these metabolites in ECF do not necessarily reflect metabolism of *releasable* dopamine, but rather intraneuronal breakdown of dopamine that need not have been

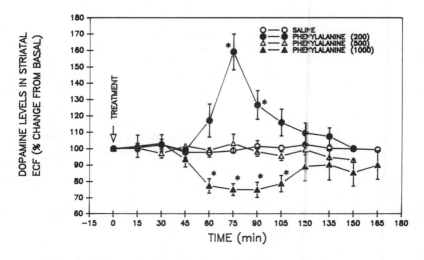

Figure 1-4. Effects of intraperitoneal administration of phenylalanine (200, 500, or 1,000 mg/kg) or saline on release of dopamine into striatal dialysate. Groups of rats ($n = 4$ for each phenylalanine dose; $n = 5$ for saline control) were anesthetized with α-chloralose/urethane (0.05/0.5 g per kg ip). After probe implantation and injury release, animals received amino acid as a suspension in saline (4 ml/kg ip) or saline alone. Samples were collected at 15-minute intervals and assayed for dopamine as described in legend to Figure 1-3 (During et al. 1988b). Vertical bars represent standard error of the mean. Statistical significance was analyzed using a one-way ANOVA for each dose; comparisons among time means performed using a least-squared means method. Asterisk denotes $P < .05$.

released. Also, there exists a variety of conditions when release and metabolism appear *uncoupled*. These findings strongly suggest that previous inferences of altered neurotransmitter release based on changes in tissue levels of dopamine metabolites may need to be challenged.

Several groups of investigators are now using brain microdialysis to study the effects of behaviors (e.g., eating, sleeping) on neurotransmitter release (Church et al. 1987; Hernandez and Hoebel 1988; Radhakishun et al. 1988), as well as the effects of pharmacological intervention on both neurotransmitter release and behaviors (Sharp et al. 1986; Zetterstrom et al. 1984). We have recently shown that awake animals that have been allowed restricted access to food show marked increases in striatal ECF levels of DOPAC, HVA, and 5-HIAA, as well as the LNAAs, with the commencement of eating. Dopamine levels failed to increase throughout this time. However, others have shown that eating increases dopamine release in both the striatum (Church et al. 1987) and the nucleus accumbens (Hernandez and Hoebel 1988; Radhakishun et al. 1988). The basis of these differences awaits clarification.

NEURONAL FIRING AND TYROSINE HYDROXYLASE ACTIVITY

The biological mechanism that couples the firing frequency of a neuron to its ability to respond to supplemental tyrosine involves phosphorylation of the enzyme tyrosine hydroxylase, a process that occurs when the firing frequency of a particular catecholaminergic neuron increases (El Mestikaway et al. 1983; Lovenberg et al. 1978). Phosphorylation of different sites on tyrosine hydroxylase can be achieved by a variety of protein kinases (Ames et al. 1978; Niggli et al. 1984). This phosphorylation, which is short lived, enhances the enzyme's affinity for its cofactor (tetrahydrobiopterin) and makes the enzyme *insensitive* to end-product inhibition (by norepinephrine and other catechols) (Ames et al. 1978; Lovenberg et al. 1975). These changes allow this enzyme's net activity to depend on the extent to which it is saturated with tyrosine.

PHYSIOLOGICAL CONSEQUENCES OF TYROSINE AVAILABILITY

If changes in tyrosine availability affect catecholamine release, then supplemental tyrosine would be expected to alter certain physiological parameters thought to involve catecholaminergic neurotransmission. Unlike the situation for tryptophan-serotonin (see Young 1985), relatively few published findings have examined and described effects of tyrosine on catecholaminergic functions.

Systemically administered tyrosine has been shown to ameliorate swim-test immobility and to increase open-field exploration in mice (Gibson et al. 1982), as well as to increase motor activity in aged mice (Thurmond and Brown 1984). Tyrosine also reverses stress-induced inhibition of open-field behavior in rats (Lehnert et al. 1984).

The coupling of tyrosine-responsiveness to neuronal firing probably explains the paradoxical effects of tyrosine on blood pressure; the amino acid *elevates* blood pressure (and sympathoadrenal catecholamine release) in hypotensive animals (Conlay et al. 1981), but *lowers* blood pressure (without effecting sympathoadrenal catecholamine release) in hypertensive animals (Sved et al. 1979b). This latter effect probably results from tyrosine's conversion to norepinephrine in brain-stem neurons active in the depressor pathway; in support of this hypothesis, tyrosine increases brain-stem MHPG-SO_4 levels in spontaneously hypertensive rats (Sved et al. 1979b). (Tyrosine fails to affect blood pressure at all in normotensive humans and animals [Glaeser et al. 1979; Sved et al. 1979b]).

Supplemental tyrosine also prevents ventricular arrhythmias in dogs (Scott et al. 1981), reverses renal hypertension in rats (Breshnahan et al. 1980), and restores estrous cycling in aged anestrous female rats (Linnoila and Cooper 1976).

Clinically, supplemental tyrosine may be useful for treating some patients with early Parkinson's disease (Growdon 1979), although its effect apparently is short lived. It may also have some use in depression, given with or without 5-hydroxytryptophan or tryptophan (Growdon 1979; Mouret et al. 1988; van Praag and Lemus 1985). Its utility in treating hypertension or other cardiovascular diseases (e.g., cardiac arrhythmias) awaits evaluation. The amino acid may also have some value in prophylaxis or treatment of stress responses; rats subjected to tail-shock stress were found, immediately thereafter, to have depressed brain norepinephrine levels, particularly in the locus coeruleus and the hypothalamus, probably reflecting the inability of norepinephrine synthesis to keep up with its release (Lehnert et al. 1984); the animals also showed norepinephrine-related behavioral abnormalities and elevated plasma corticosterone levels (Reinstein et al. 1985). All of these changes, including the adrenocortical response, were suppressed by supplemental orally administered tyrosine, but not if the tyrosine was coadministered with another LNAA (valine) that blocked its brain uptake.

In a preliminary study, Bandaret and Lieberman (1989) investigated the effect of tyrosine on performance, symptoms, and mood of United States Air Force pilot volunteers under situations of stress (i.e., a cold environment at low barometric pressure). Tyrosine enhanced performance, including reaction time, vigilance, and complex information processing, and reduced subjective symptoms of cold, muscle discomfort, and headache. Mood states (e.g., anxiety and tension) were also improved.

Precursor control of catecholamine release may now be of particular relevance because of the consumption of the artificial dipeptide sweetener aspartame (56% phenylalanine by weight). If our microdialysis data showing that consumption of phenylalanine can alter dopamine release in rodent brain is also true for humans, then this factor may be involved in the various neurological symptoms sometimes described as being temporally associated with aspartame consumption, for example, seizures (Maher and Pinto 1988) and precipitation of headaches (Johns 1988).

The competitive nature of brain LNAA uptake may underlie variations in the therapeutic effect of L-dopa, itself an LNAA. The "on-off" effect in parkinsonian patients receiving L-dopa is significantly worsened by ingestion of a high-protein meal (Nutt et al. 1984). A high-carbohydrate meal, which elevates the plasma L-dopa ratio by lowering plasma LNAA levels, exacerbated dyskinesias after L-dopa administration; a high-protein meal, which raises plasma LNAA levels, exacerbated the parkinsonian symptoms (Wurtman et al. 1988).

CONCLUSIONS

Tyrosine administration increases, sequentially, the "tyrosine ratio" and brain tyrosine levels. With catecholamine neurons, this increase can facilitate synthesis and release of the catecholamine neurotransmitters. In otherwise untreated animals, systemic tyrosine administration causes a transient increase in dopamine release from nigrostriatal neurons. This increase can be enhanced and prolonged by treatments that interfere with multisynaptic or autoreceptor-mediated feedback processes. If animals are subjected to treatments that increase the firing frequency of particular catecholaminergic tracts or nerves, the neurons remain sensitive to changes in the available tyrosine level. The possible use of supplemental tyrosine to treat neurological and behavioral disorders is under examination.

REFERENCES

Acworth IN, Nicholass J, Morgan B, et al: Effect of sustained exercise on concentrations of plasma aromatic and branched-chain amino acids and brain amines. Biochem Biophys Res Commun 137:149–153, 1986

Acworth IN, During MJ, Wurtman RJ: Tyrosine: effects on catecholamine release. Brain Res Bull 21:473–477, 1988

Agid Y, Javoy F, Glowinski J: Hyperactivity of the remaining dopaminergic neurons after partial destruction of the nigrostriatal dopaminergic system of the rat. Nature New Biol 245:150–151, 1973

Ames MM, Lerner P, Lovenberg W: Tyrosine hydroxylase: activation by protein phosphorylation and product inhibition. J Biol Chem 253:27–31, 1978

Anden NE, Grabowska-Anden M, Lindgren S, et al: Synthesis rate of dopamine: difference between corpus striatum and limbic system as a possible explanation in variations in reactions to drugs. Naunyn Schmiedebergs Arch Pharmacol 323:193–198, 1983

Annunziato L: Regulation of the tuberoinfundibular and nigrostriatal systems. Neuroendocrinology 29:66–76, 1979

Annunziato L, Moore KE: Prolactin in CSF selectively increases dopamine turnover in the median eminence. Life Sci 22:2037–2042, 1978

Badawy AA-B, Evans M: Inhibition of rat liver pyrrolase activity and elevation of brain tryptophan by administration of antidepressants. Biochem Pharmacol 30:1211–1216, 1981

Badawy AA-B, Williams DL: Enhancement of rat brain catecholamine synthesis by administration of small doses of tyrosine and evidence for substrate inhibition of tyrosine hydroxylase activity by large doses of the amino acid. Biochem J 206:165–168, 1982

Banderet LE, Lieberman HR: Treatment with tyrosine, a neurotransmitter precursor, reduces environmental stress in humans. British Research Bulletin 22:759–762, 1989

Beal MF, Martin JB: Topographic dopamine and serotonin distribution and turnover in rat striatum. Brain Res 358:10–15, 1985

Bradford HF: Chemical Neurobiology. New York, WH Freeman, 1986

Breshnahan MR, Hatzinikolaou P, Brunner HR, et al: Effects of tyrosine infusion in normotensive and hypertensive rats. Am J Physiol 239:H206–H211, 1980

Bunney BS, Walters JR, Roth RH, et al: Dopaminergic neurons: effect of antipsychotic drugs and amphetamine on single cell activity. J Pharmacol Exp Ther 185:560–571, 1973

Carlsson A, Lindqvist M: Dependents of 5-HT and catecholamine synthesis on precursor amino acid levels in rat brain. Naunyn Schmiedebergs Arch Pharmacol 303:157–164, 1978

Church WH, Justice JB, Neill DB: Detecting behaviorally relevant changes in extracellular dopamine with microdialysis. Brain Res 412:397–399, 1987

Cohen EL, Wurtman RJ: Brain acetylcholine: increase after systemic choline administration. Life Sci 16:1095–1102, 1975

Conlay LA, Maher TJ, Wurtman RJ: Tyrosine increases blood pressure in hypotensive rats. Science 212:559–560, 1981

Conlay LA, Wurtman RJ, Lopez G, et al: Effects of running the Boston marathon on plasma concentrations of large neutral amino acids. J Neural Transm 76:65–71, 1989

Curzon G: Transmitter metabolism and behavioral abnormalities in liver failure, in The Biochemistry of Psychiatric Disturbances. Edited by Curzon G. Chichester, UK, John Wiley, 1980, pp 89–111

Curzon G, Kantamaneni BD, Callighan N, et al: Brain transmitter precursors and metabolites in diabetic ketoacidosis. J Neurol Neurosurg Psychiatry 45:489–493, 1982

Decombaz J, Reinhardt P, Anantharaman K, et al: Biochemical changes in a 100 km run: free amino acids, urea and creatine. Eur J Appl Physiol 41:61–72, 1979

During MJ, Acworth IN, Wurtman RJ: Effects of systemic L-tyrosine on dopamine release from the rat corpus striatum and nucleus accumbens. Brain Res 452:378–380, 1988a

During MJ, Acworth IN, Wurtman RJ: Phenylalanine administration influences dopamine release in the rat's corpus striatum. Neurosci Lett 93:91–95, 1988b

During MJ, Acworth IN, Wurtman RJ: Dopamine release in rat striatum: physiological coupling to tyrosine supply. J Neurochem 52:1449–1454, 1989

El Mestikaway S, Glowinski J, Hamon M: Tyrosine hydroxylase activation in depolarized dopaminergic terminals: involvement of calcium-dependent phosphorylation. Nature 302:830–832, 1983

Fernstrom JD, Faller DV: Neutral amino acids in the brain: changes in a response to food ingestion. J Neurochem 30:1531–1538, 1978

Fernstrom JD, Wurtman RJ: Brain serotonin content: physiological dependence on plasma tryptophan levels. Science 173:149–152, 1971a

Fernstrom JD, Wurtman RJ: Brain serotonin content: increase following ingestion of carbohydrate diet. Science 174:1023–1025, 1971b

Fernstrom JD, Wurtman RJ: Brain serotonin content: physiological regulation by plasma neutral amino acids. Science 178:414–416, 1972a

Fernstrom JD, Wurtman RJ: Elevation of plasma tryptophan by insulin in the rat. Metabolism 21:337–342, 1972b

Fernstrom JD, Wurtman RJ, Hammarstrom-Wiklund B, et al: Diurnal variations in the plasma concentrations of tryptophan, tyrosine and other large neutral amino acids: effect of dietary protein intake. Am J Clin Nutr 32:1912–1922, 1979

Fuller RW, Snoddy HD: L-Tyrosine enhancement of the elevation of 3,4-dihydroxyphenylacetic acid concentration by spiperone and amfonelic acid. J Pharm Pharmacol 34:117–118, 1982

Gibson CJ: Factors controlling brain catecholamine biosynthesis: effect of brain tyrosine. Unpublished doctoral dissertation, Massachusetts Institute of Technology, Cambridge, MA, 1977

Gibson CJ: Dietary control of retinal dopamine synthesis. Brain Res 382:195–198, 1986

Gibson CJ: Alterations in retinal tyrosine and dopamine levels in rats consuming protein or tyrosine-supplemented diets. J Neurochem 50:1769–1774, 1988

Gibson CJ, Wurtman RJ: Physiological control of brain norepinephrine synthesis by brain tyrosine concentration. Life Sci 22:1399–1406, 1978

Gibson CJ, Deikel SM, Young SN, et al: Behavioral and biochemical effects of tryptophan, tyrosine and phenylalanine in mice. Psychopharmacology (Berlin) 76:118–121, 1982

Gibson CJ, Watkins CJ, Wurtman RJ: Tyrosine administration enhances dopamine synthesis and release in light-activated retinas. J Neural Transm 56:153–160, 1983

Glaeser B, Melamed E, Growdon JH, et al: Elevation of plasma tyrosine after a single oral dose of L-tyrosine. Life Sci 25:265–272, 1979

Growdon JH: Neurotransmitter precursors in the diet: their use in the treatment of brain diseases, in Nutrition and the Brain, Vol 3. Edited by Wurtman RJ, Wurtman JJ. New York, Raven Press, 1979, pp 117–181

Guroff G, King W, Udenfriend S: The uptake of tyrosine by rat brain in vitro. J Biol Chem 236:1773–1777, 1961

Haubrich DR, Wang PF, Herman RL, et al: Acetylcholine synthesis in rat brain: dissimilar effects of clozapine and chlorpromazine. Life Sci 17:739–748, 1975

Hernandez L, Hoebel BG: Food reward and cocaine increase extracellular dopamine in the nucleus accumbens as measured by microdialysis. Life Sci 42:1705–1712, 1988

Ikeda M, Levitt M, Udenfriend S: Phenylalanine as substrate and inhibitor of tyrosine hydroxylase. Arch Biochem Biophys 120:420–427, 1967

Imperato A, Di Chiara G: Trans-striatal dialysis coupled to reverse phase high performance liquid chromatography with electrochemical detection: a new method for the study of in vivo release of endogenous dopamine and metabolites. J Neurosci 4:966–977, 1984

Irie K, Wurtman RJ: Release of norepinephrine from rat hypothalamic slices: effects of desipramine and tyrosine. Brain Res 423:391–394, 1987

Johns DR: Aspartame and headache, in Dietary Phenylalanine and Brain Function. Edited by Wurtman RJ, Ritter-Walker E. Boston, MA, Birkhauser, 1988, pp 303–312

Katz I, Lloyd T, Kaufman S: Studies on phenylalanine and tyrosine hydroxylation by rat brain tyrosine hydroxylase. Biochim Biophys Acta 445:567–578, 1976

Knott PJ, Curzon G: Free tryptophan in plasma and brain tryptophan. Nature 239:452–453, 1972

Lehnert H, Reinstein DK, Strowbridge BW, et al: Neurochemical and behavioral consequences of acute, uncontrollable stress: effects of dietary tyrosine. Brain Res 303:2157–2163, 1984

Liebschutz J, Airoldi L, Brownstein MJ, et al: Regional distribution of endogenous and parenteral glutamate, aspartate and glutamine in rat brain. Biochem Pharmacol 26:443–446, 1977

Linnoila M, Cooper RL: Reinstatement of vaginal cycles in aged female rats. J Pharmacol Exp Ther 199:477–482, 1976

Lovenberg W, Bruckwick EA, Hanbauer I: ATP, cyclic-AMP, and magnesium increase the affinity of rat striatal tyrosine hydroxylase for its cofactor. Proc Natl Acad Sci USA 72:2955–2958, 1975

Lovenberg W, Ames MM, Lerner P: Mechanisms of short-term regulation of tyrosine hydroxylase, in Psychopharmacology: A Generation of Progress. Edited by Lipton MA, DiMascio A, Killam KF. New York, Raven Press, 1978, pp 247–259

Madras BK, Cohen EL, Messing R, et al: Relevance of serum free tryptophan to tissue tryptophan concentrations. Metabolism 23:1107–1116, 1974

Maher TJ, Pinto JMB: Aspartame, phenylalanine, and seizures in experimental animals, in Dietary Phenylalanine and Brain Function. Edited by

Wurtman RJ, Ritter-Walker E. Boston, MA, Birkhauser, 1988, pp 95–103

Maher TJ, Wurtman RJ: L-Threonine administration increases glycine concentration in the rat central nervous system. Life Sci 26:1283–1286, 1980

Maher TJ, Glaeser BS, Wurtman RJ: Diurnal variations in plasma concentrations of basic and neutral amino acids in red cell concentrations of aspartate and glutamate: effects of dietary protein intake. Am J Clin Nutr 39:722–729, 1984

Maire J-CE, Wurtman RJ: Choline production from choline-containing phospholipids: a hypothetical role in Alzheimer's disease and aging. Neuro-Psychol Biol Psychiatry 8:637–642, 1984

Marsden CA: Measurement of Neurotransmitter Release in Vivo. New York, John Wiley, 1984

McMenamy RH, Oncley JL: The specific binding of L-tryptophan to serum albumin. J Biol Chem 233:1436–1440, 1958

Melamed E, Hefti F, Wurtman RJ: Tyrosine administration increases striatal dopamine release in rats with partial nigrostriatal lesions. Proc Natl Acad Sci USA 77:4305–4309, 1980

Milner JD, Wurtman RJ: Tyrosine availability determines stimulus-evoked dopamine release from rat striatal slices. Neurosci Lett 59:215–220, 1985

Milner JD, Wurtman RJ: Catecholamine synthesis: physiological coupling to precursor supply. Biochem Pharmacol 35:875–881, 1986

Milner JD, Irie K, Wurtman RJ: Effects of phenylalanine on the endogenous release of dopamine from rat striatal slices. J Neurochem 47:1444–1448, 1986

Milner JD, Reinstein DK, Wurtman RJ: Dopamine synthesis in rat striatum: mobilization of tyrosine from non-dopaminergic cells. Experientia 43:1109–1110, 1987

Moldawer LL, Kamamura I, Bistrian RR, et al: The contribution of phenylalanine to tyrosine metabolism in vivo. Studies in the postabsorptive and phenylalanine-loaded rat. Biochem J 210:811–817, 1983

Moore KE, Annunziato L, Gudelsky GA: Studies on the tuberoinfundibular dopamine neurons. Adv Biochem Psychopharmacol 19:193–204, 1978

Morgenroth VH, Walters JR, Roth RH: Dopaminergic neurons: alterations in the kinetic properties of tyrosine hydroxylase after cessation of impulse flow. Biochem Pharmacol 26:655–661, 1976

Mouret J, Lemoine P, Minuit M-P, et al: Immediate and long lasting treatment of dopamine dependent depression (DDD) by L-tyrosine: a clinical and polygraphic study. CR Acad Sci Paris 306:93–98, 1988

Niggli V, Knight DE, Baker PF, et al: Tyrosine hydroxylase in "leaky" adrenal medullary cells: evidence for in situ phosphorylation by separate calcium and cyclic-AMP dependent systems. J Neurochem 43:646–658, 1984

Nutt JG, Woodward WR, Hammarstad JP, et al: The "on-off" phenomenon in Parkinson's disease: relation to levodopa absorption and transport. N Engl J Med 310:483–488, 1984

Oldendorf WH: Brain uptake of radiolabeled amino acids, amines and hexoses after arterial injection. Am J Physiol 221:1629–1639, 1971

Pardridge WM: Regulation of amino acid availability to the brain, in Nutrition and the Brain, Vol 1. Edited by Wurtman RJ, Wurtman JJ. New York, Raven Press, 1977, pp 141–204

Pardridge WM: Potential effects of the dipeptide sweetener aspartame on the brain, in Nutrition and the Brain, Vol 7. Edited by Wurtman RJ, Wurtman JJ. New York, Raven Press, 1986, pp 199–241

Pardridge WM, Oldendorf WH: Kinetic analysis of blood-brain barrier transport of amino acids. Biochim Biophys Acta 401:128–136, 1986

Radhakishun FS, van Ree JM, Westerink BHC: Scheduled eating increases dopamine release in the nucleus accumbens of food-deprived rats as assessed with on-line brain microdialysis. Neurosci Lett 85:351–356, 1988

Reinstein DK, Lehnert H, Wurtman RJ: Tyrosine prevents behavioral and neurochemical correlates of an acute stress in rats. Life Sci 34:2225–2232, 1984

Reinstein DK, Lehnert H, Wurtman RJ: Dietary tyrosine suppresses the rise in plasma corticosterone following acute stress in rats. Life Sci 37:2157–2163, 1985

Scally MC, Ulus I, Wurtman RJ: Brain tyrosine level controls striatal dopamine synthesis in haloperidol-treated rats. J Neural Transm 41:1–6, 1977

Schwarz JC, Lampart C, Rose C: Histamine formation in rat brain in vivo: effects of histidine loads. J Neurochem 19:801–810, 1972

Scott NA, DeSilva RA, Lown B, et al: Tyrosine administration decreases vulnerability to ventricular fibrillation in the normal canine heart. Science 211:727–729, 1981

Sharp T, Ljungberg T, Zetterstrom T, et al: Intracerebral dialysis coupled to a novel activity box: a method to monitor dopamine release during behavior. Pharmacol Biochem Behav 24:1755–1759, 1986

Shiman R, Akino M, Kaufman S: Solubulization and partial purification of tyrosine hydroxylase from bovine adrenal medulla. J Biol Chem 246:1330–1340, 1971

Stanbury JB, Wyngaardan JB, Fredrickson DS: The Metabolic Basis For Inherited Diseases. New York, McGraw-Hill, 1966

Sved AF: The relationship between tyrosine availability and catecholaminergic function: physiological and biochemical studies. Unpublished doctoral dissertation, Massachusetts Institute of Technology, Cambridge, MA, 1980

Sved AF: Precursor control of the function of monoaminergic neurons, in Nutrition and the Brain, Vol 6. Edited by Wurtman RJ, Wurtman JJ, New York, Raven Press, 1983, pp 223–275

Sved A, Fernstrom JD: Tyrosine availability and dopamine synthesis in the striatum: studies with gamma-butyrolactone. Life Sci 29:743–748, 1981

Sved AF, Fernstrom JD, Wurtman RJ: Tyrosine administration decreases serum prolactin levels in chronically reserpinized rats. Life Sci 25:1293–1300, 1979a

Sved AF, Fernstrom JD, Wurtman RJ: Tyrosine administration reduces blood pressure and enhances brain norepinephrine release in spontaneously-hypertensive rats. Proc Natl Acad Sci USA 76:3511–3514, 1979b

Tam SY, Roth RH: Tyrosine preferentially increases dopamine synthesis and release in mesocortical dopamine neurons with high firing frequency. Neurosci Abstr 10:881, 1984

Thurmond JB, Brown JW: Effect of brain monoamine precursors on stress induced behavioral and neurochemical changes in aged mice. Brain Res 296:93–102, 1984

Ungerstedt U: Measurement of neurotransmitter release by intracranial dialysis, in Measurement of Neurotransmitter Release In Vivo. Edited by Marsden CA. New York, John Wiley, 1984, pp 81–105

Valzelli L, Bernasconi S, Cohen E, et al: Effect of different psychoactive drugs on serum and brain tryptophan levels. Neuropsychobiology 6:224–229, 1980

van Praag HM, Lemus C: Monoamine precursors in the treatment of psychiatric disorders, in Nutrition and the Brain, Vol 7. Edited by Wurtman RJ, Wurtman JJ. New York, Raven Press, 1985, pp 89–138

Walters JR, Roth RH: Dopaminergic neurons: drug-induced antagonism of the increase in tyrosine hydroxylase activity produced by cessation of impulse flow. J Pharmacol Exp Ther 191:82–91, 1974

Walters JR, Roth RH: Dopaminergic neurons: an in vivo system for measuring drug interactions with presynaptic receptors. Naunyn Schmiedebergs Arch Pharmacol 296:5–14, 1976

Walters JR, Roth R, Aghajanian G: Dopaminergic neurons: similar biochemical and histochemical effects of hydroxybutyrate and acute lesions of the nigroneostriatal pathway. J Pharmacol Exp Ther 186:630–639, 1973

Westerink BHC, Wirix EJ: On the significance of tyrosine for the synthesis and catabolism of dopamine in the rat brain: evaluation by HPLC with electrochemical detection. J Neurochem 40:758–764, 1983

Wood JL, Allison RG: Effects of consumption of choline and lecithin on neurological and cardiovascular systems, in FASEB: Technical Report. Washington, DC, SRO Press, 1981, pp 1–105

Wurtman RJ, Caballero B, Salzman E: Facilitation of L-dopa-induced dyskinesias by dietary carbohydrate. N Engl J Med 319:1288–1289, 1988

Wurtman RJ, Larin S, Mostafapour S, et al: Brain catechol synthesis: control by brain tyrosine concentrations. Science 185:183–184, 1974

Yamori Y, Fujiwara M, Horie R, et al: The hypotensive effect of centrally administered tyrosine. Eur J Pharmacol 68:201–204, 1980

Young S: The clinical psychopharmacology of tryptophan, in Nutrition and the Brain, Vol 7. Edited by Wurtman RJ, Wurtman JJ. New York, Raven Press, 1985, pp 49–88

Yuwiler A, Oldendorf WH, Geller E, et al: The effect of albumin binding and amino acid competition on tryptophan uptake into the brain. J Neurochem 28:1015–1023, 1977

Zetterstrom T, Sharp T, Ungerstedt U: Effect of neuroleptic drugs on striatal dopamine release and metabolism in the awake rat studied by intracerebral dialysis. Eur J Pharmacol 106:27–37, 1984

Zetterstrom T, Herrera-Marschitz M, Ungerstedt U: Simultaneous measurement of dopamine release and rotational behavior in 6-hydroxydopamine denervated rats using intracerebral dialysis. Brain Res 376:1–7, 1986

Chapter 2

Availability of Amino Acids to the Brain and Implication for Transmitter Amine Function

Gerald Curzon, Ph.D., D.Sc.

Chapter 2

Availability of Amino Acids to the Brain and Implication for Transmitter Amine Function

This chapter is a discussion of how physiological variables (e.g., feeding, stress, exercise) affect brain amino acid concentrations. The focus is mainly on amino acids that are precursors of transmitter amines, and especially on tyrosine and tryptophan, which tyrosine hydroxylase and tryptophan hydroxylase, respectively, convert to 3-(3,4-dihydroxyphenyl)alanine (dopa) and 5-hydroxytryptophan by reactions that are rate limiting for catecholamine and 5-hydroxytryptamine (serotonin [5-HT]) synthesis. Precursor loading experiments suggest that brain tyrosine hydroxylase is normally close to saturation with its substrate (at least when catecholamine neurons are quiescent) and is therefore relatively insensitive to altered substrate availability, but that tryptophan hydroxylase is only about 50% saturated and therefore responds to substrate changes more sensitively (Carlsson and Lindqvist 1978; Sved 1983). Far less attention has been paid to the fact that histidine decarboxylase is normally far below saturation with its precursor amino acid (Pardridge 1986), and, thus, the synthesis of its metabolite, histamine, is far more responsive to changes of precursor availability than that of 5-HT or the catecholamines.

Before the important studies of Fernstrom and Wurtman (1971, 1973), there was what now seems a surprising lack of interest in the possibility that the ranges of tryptophan and tyrosine availabilities to the brain that were apparent under physiological conditions might affect the synthesis of transmitter amines. Their work was a major stimulant of almost two decades of intense and valuable activity on the transport of tryptophan and tyrosine to the brain. However, accumulating data indicate that, in general, a number of powerful influences tend to maintain the stability of brain amino acid pools.

Indeed, despite the recent considerable interest in the idea that diet-induced changes of 5-HT synthesis control appetite, many findings suggest that the effects of diet (on brain tryptophan at least) normally serve mainly to ensure sufficient supplies of tryptophan for central 5-HT synthesis. However, as other evidence points to large interindividual differences in the responsiveness of transmitter synthesis to precursor supply, there may be subgroups in which serotonergic functional activity responds directly to feeding. Also, neuronal activity may affect relationships between precursor availability and transmitter synthesis. This chapter contains some comments on these possibilities and on how dietary supply, stress, and exercise affect the availability of amino acids to the brain, especially those that are transmitter precursors (in particular, tryptophan). It also contains some remarks on possible behavioral implications.

FEEDING

The stability of brain amino acid pools is, in part, a consequence of control mechanisms that tend to preserve plasma amino acid levels. For example, even fasting for up to 5 days has surprisingly little effect on plasma concentrations of most amino acids in the rat, apart from phenylalanine, which increases somewhat, and alanine, arginine, glutamate, and threonine, which show moderate decreases (Enwonowu 1987). Brain amino acid values, on the whole, do not alter significantly, and some of the changes that do occur (Knott et al. 1973) are readily explicable in terms of the plasma findings reported by Enwonowu; that is, the increase in levels of brain phenylalanine and the decrease in levels of brain arginine and alanine. Plasma glutamate values decrease, but this factor would not be expected to alter brain values, because the uptake site is normally saturated (Olendorf and Szabo 1976; Pardridge 1979a).

Two amino acids, however, show marked and highly significant discrepancies between the effects of fasting on their plasma and brain levels: threonine, despite a considerable fall in the plasma on 24-hour starvation (Enwonowu 1987), shows values that remain unaltered in the brain; tryptophan, despite unaltered total plasma levels (i.e., free tryptophan plus tryptophan loosely bound to albumin), increases in the brain (Curzon et al. 1972). It is also superficially anomalous that brain tryptophan values increase if rats are given a large tryptophan-free high carbohydrate meal (Fernstrom and Wurtman 1971; Sarna et al. 1984).

Thus, brain tryptophan levels increase in two circumstances in which decreases might have been expected. These results are largely explicable as follows:

1. Fasting increases plasma levels of unesterified fatty acids that displace tryptophan from albumin, so that the amino acid becomes more available to the brain (Curzon et al. 1973; Knott and Curzon 1972).
2. A high-carbohydrate meal causes insulin secretion, with resultant increased amino acid uptake by muscle, so that plasma levels of the large neutral amino acids (LNAAs) that compete with tryptophan for transport to the brain decrease.

As a result, the ratio of plasma tryptophan values to those of competers rises (Fernstrom and Wurtman 1973). Unfortunately, there was much confusion in the past about the evidence for these mechanisms, since many of the earlier papers on the effects of feeding on brain tryptophan levels did not include data on all the relevant variables. Also, as discussed previously (Curzon 1979), some investigations suffered from methodological defects. As a result, the relative importance of the above mechanisms for the increase in brain tryptophan levels was often disputed. However, detailed analyses suggest that in both rats and humans (Perez-Cruet et al. 1974; Sarna et al. 1984), both mechanisms can be involved in the effects of feeding on brain tryptophan.

How important are the mechanisms? An investigation of psychiatric patients undergoing subcaudate tractotomy showed that the variation of cortical and ventricular tryptophan values within the group correlated well with plasma free tryptophan values (Gillman et al. 1981) but that correlations with plasma total tryptophan were weaker and, in general, correlations became weaker if the plasma tryptophan values were (as in the rat experiments [Fernstrom et al. 1973]) divided by the sum of the plasma concentrations of the LNAAs. The LNAA values in this study varied over a 1.5-fold range, which implies that moderate plasma LNAA changes may have no more than minor effects on tryptophan transport to the human brain. This conclusion agrees well with calculations from influx experiments using rats (Curzon 1985; Mans et al. 1979).

The above human study was done on anesthetized subjects who had not eaten for 14 hours. Nevertheless, when considering the role of dietary effects on plasma LNAA levels in the control of brain tryptophan (and other amino acids), it is pertinent that total plasma LNAA concentrations in a group of freely feeding adults following their own dietary preference were essentially unaffected by breakfast and rose only about 15% above fasting values after lunch (Scriver et al. 1985). Similarly, other workers found no significant diurnal variation of plasma amino acids in human subjects under normal British dietary conditions (Milson et al. 1979). Even when meals of considerably different composition were given to humans, plasma tyrosine and tryptophan concentrations varied approximately in proportion to the values for other LNAAs. Thus, plasma competer ratios of subjects given carbohydrate or 20% protein meals (Ashley et al. 1982, 1985) altered moderately, if at all. The reported changes would be expected to have little, if any, effect on the availability of 5-HT to receptors.

These findings do not contradict the large body of evidence that the *presence* of LNAAs in plasma alters the availability of tryptophan and other individual LNAAs to the brain and that certain extremes of meal composition can have substantial effects on concentrations of amino acids therein (Glaeser et al. 1983; Møller 1985; Yokohoshi and Wurtman 1986). However, they indicate that normal acute dietary variations of the ratio of plasma tryptophan to LNAAs probably have rather small effects. Indeed, in the above studies of Ashley et al. (1982, 1985), the ratio increased considerably only when a tryptophan supplement was added to the meal.

Even a reported appreciable change in the ratio must be interpreted with caution. In human studies, with the exception of the work of Gillman et al. (1981), plasma and brain data have (for obvious reasons) rarely been obtainable on the same subjects. Therefore, ratios of plasma tryptophan to LNAAs are not infrequently used as the sole indices of brain tryptophan level. In some studies, the ratio changes simply because plasma tryptophan concentration has changed; in others the ratio is reported without data on its numerator and denominator. It is relevant here that, in a number of nondietary studies, correlations between plasma and brain tryptophan values were not strengthened but weakened by taking competing LNAAs into account (Bloxam and Curzon 1978; Gillman et al. 1981; Sarna et al. 1982). Also, a recent study (Fernstrom et al. 1987) indicated that physiologically relevant differences in chronic protein intake did not cause dose-related changes in the ratio of plasma tryptophan to LNAAs. Furthermore, in these experiments, the ratios did not, on the whole, predict brain values. Notwithstanding these indications that the brain concentrations of amino acids which are transmitter precursors do not markedly depend on acute changes of dietary intake, a note of caution is needed as the conclusion is based on investigations of relatively small numbers of subjects over short periods. As values obviously range more widely within larger groups or when individuals are studied over longer periods, it is reasonable to expect the existence of subgroups whose central amino acid concentrations are vulnerable to dietary influence. Such subgroups are likely to be revealed when an amino acid is given as a drug to large numbers of subjects and its therapeutic effect depends on its transport to the brain. The dopa treatment of Parkinson's disease provides an example, as a low-protein diet increased the "on" time, (i.e., the periods over which a dose of the drug was effective) in patients who suffered from the "on-off" effect. This was explained by the decreased plasma concentrations of LNAAs that compete with the transport of L-dopa to the brain (Eriksson et al. 1988).

STRESS

Early work (Bliss et al. 1968) showed that brain 5-HT metabolism increased on 2-hour immobilization stress. This increase was subsequently shown to be associated with a rise of brain tryptophan (Curzon et al. 1972). Initially, the latter effect was attributed to the associated increase of plasma free tryptophan (Curzon and Knott 1974; Knott and Curzon 1972) that resulted from a stress-induced rise of plasma nonesterified fatty acids. However, in a study (Kennett et al. 1986) described below, the effect was more convincingly explained as part of a general alteration of the plasma-brain relationship for LNAAs. The investigation showed that immobilization decreased plasma amino acid values by 20 to 50%. The only exception was free tryptophan, which, as before, rose while total tryptophan fell (in common with 12 of 15 of the other amino acids measured). Other workers (Milakofsky et al. 1985) obtained similar but less marked changes on briefer immobilization. The decrease could be due to stress-provoked increases in plasma catecholamines, as isoprenaline infusion increases plasma free tryptophan but decreases total tryptophan, and tyrosine (Hutson et al. 1980), and adrenaline is reported to decrease most amino acids in human plasma (Shamoon et al. 1980).

The plasma changes on immobilization (Kennett et al. 1986) were not associated with parallel changes in the brain. On the contrary, brain tryptophan and four other LNAAs (phenylalanine, valine, leucine, and isoleucine) rose significantly, and no amino acid fell significantly. The increases were not explicable by brain influx rates, calculated using plasma amino acid values and published kinetic data (Pardridge 1977, 1979a). This is strikingly shown in Figure 2-1, as (with the possible exception of histidine) brain LNAA values of immobilized rats are higher than predictable from calculated influxes. Thus, immobilization appears to alter relationships between plasma and brain LNAA concentrations in favor of the brain. A similar shift occurs on portocaval anastomosis (Bloxam and Curzon 1978; Mans et al. 1984).

The above effects of immobilization could be due to breakdown of the blood-brain barrier as indicated by the penetration of albumin

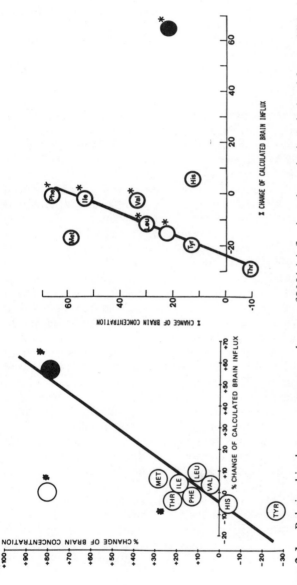

Figure 2-1. Relationships between percentage changes of LNAA influx into the rat brain (calculated from published K_m and V_{max} values [Pardridge 1977]) and percentage increases of brain LNAA. The effects of 2-hour trained running are shown on the left; those of 2-hour immobilization are shown on the right. Standard abbreviations are used for amino acids except that ○ and ● refer to plasma total and free tryptophan, respectively. Significant increases of concentrations upon running or immobilization are indicated by asterisks. The regression line for the effect of exercise was calculated using influx values for all the LNAAs, including plasma free (but not total) tryptophan ($r = 0.93$, $n = 9$, $P < .001$). The line for the effect of immobilization was calculated using influx values for seven of the LNAAs (including all five for which the increase in brain concentration was significant) and data for plasma *total* (but not free) tryptophan ($r = 0.94$, $n = 7$, $P < .01$). Results from Chaouloff et al. 1985a; Kennett et al. 1986.

into the brain during immobilization (Belova and Jonsson 1982). However, this breakdown was detected in very few brain regions and thus does not explain the amino acid changes, which were determined on the whole brain (Kennett et al. 1986). Other mechanisms that cannot be excluded could involve altered brain amino acid metabolism or efflux. Changes in cerebral blood flow are not likely to be responsible, as, at most, they only moderately alter amino acid influx (Hawkins et al. 1982), and immobilizing conscious normal rats only slightly affects flow (Ohata et al. 1981). It seems more likely that transport kinetics across the barrier are altered. This alteration could be due to stress-provoked hormonal changes. For example, thyroid hormones, whose plasma levels are initially increased in stress (Langer et al. 1983), increase amino acid transport to the brain (Daniel et al. 1975). The enhancement of brain uptake of LNAAs by isoprenaline (Eriksson and Carlsson 1982; Hutson et al. 1980) suggests that increased catecholamine secretion could be involved. Plasma growth hormone increases in stress (Kant et al. 1983), but data on its effect on transport are equivocal (Cocchi et al. 1975; Tang and Cotzias 1976).

Kennett et al. (1986) found that plasma free tryptophan and brain tryptophan rose in the same proportion. Nevertheless, the relationship between the effects of immobilization on brain tryptophan concentration and influx is highly consistent with the relationships for other LNAAs if plasma total tryptophan is used to calculate tryptophan influx but not if plasma free tryptophan is used (Figure 2-1). This could be due to an effect of immobilization on the extraction of plasma tryptophan by the brain so that the transport site competes effectively with plasma albumin for the amino acid, stripping it off from the protein (Pardridge 1979b); thus, the brain "sees" all the plasma tryptophan instead of only the free component. Stripping off would be facilitated by the decreased K_m values for uptake of tryptophan and other LNAAs, which are consistent with the data in Figure 2-1.

A special situation occurs when food-deprived rats are stressed. Under these circumstances, sympathetic activation of lipolysis and hence of the liberation of plasma tryptophan from its binding to albumin is more readily provoked. Thus, relatively mild stresses such as the removal of cage-mates can be sufficient to increase plasma free tryptophan and hence brain tryptophan (Knott et al. 1977).

It was remarkable that brain levels not only of LNAAs were maintained on immobilization but also those of the other amino acids, even though their plasma concentrations were decreased (Kennett et al. 1986). Results for alanine and serine may reflect the same mechanisms as those responsible for the disposition of the LNAAs, as they are at least partly transported by the same site. Results for the basic amino acids, arginine and lysine, which are transported by a different site, can be explained in terms of influx calculated as in Figure 2-1 and agree qualitatively with previous data (Banos et al. 1974). Plasma glutamate was decreased by stress (Kennett et al. 1986), but this is unlikely to affect brain values, as the uptake is normally saturated (Pardridge 1977).

Results, in general, reveal that different mechanisms ensure that the general decrease of plasma amino acid concentrations on immobilization for 2 hours does not lead to decreases in the brain values. Instead, brain values are maintained or increased. The increase of brain tryptophan seems to be at least partially responsible for the increased brain 5-HT metabolism that occurs not only upon immobilization (Curzon et al. 1972; Knott and Curzon 1972) but also upon foot shock stress (Dunn 1988). The maintenance of brain tyrosine levels despite a fall of plasma tyrosine is also of interest, as immobilization alters the relationship between its level in the brain and dopamine synthesis (Marcou et al. 1987).

EXERCISE

Running substantially increases brain and cerebrospinal fluid (CSF) tryptophan concentration in rats trained to run on a treadmill. Plasma free (but not total) tryptophan rose (Chaouloff et al. 1985a, 1986a) because of exercise-induced lipolysis, and the correlation between free and brain tryptophan concentrations suggests a causal relationship (Chaouloff et al. 1985a). Despite the evidence against explaining the immobilization stress-induced brain tryptophan increases in this way (see above), investigation (Chaouloff et al. 1986b) confirmed that the rise of brain tryptophan upon exercise could only be explained in terms of the rise of plasma free tryptophan. Running for 2 hours at 20 m/minute almost doubled both plasma free and brain values. Although increased arterial concentrations of LNAAs are reported following exercise in humans (Ahlborg et al. 1974; Felig and Wahren 1971), other rat plasma LNAAs were essentially unaffected.

Exercise increased only one other rat brain LNAA (threonine). This finding was not explicable in terms of the plasma values and is of interest because a similar threonine-specific discrepancy between plasma and brain changes was noted upon fasting (Enwonowu 1987). Exercise also increased the basic amino acid lysine in the brain but decreased glycine, alanine, and γ-aminobutyric acid (GABA). Brain tryptophan could conceivably have increased because of hyperammonemia, as this effect is reported to occur on prolonged exercise

(Mutch and Banister 1983) and could lead to increased brain tryptophan (Chaouloff et al. 1985b) as the result of a rise of brain glutamine. However, this mechanism seems improbable in the study of Chaouloff et al. (1986b), because the transport of other LNAAs to the brain would also be affected (Rigotti et al. 1985), and this did not occur.

The different mechanisms by which exercise and immobilization influence brain amino acids may reflect greater stress and resultant greater sympatho-adrenal stimulation in the latter group (Figure 2-2) so that brain uptake of LNAAs increases (Eriksson and Carlsson 1982; Hutson et al. 1980). Thus, exercise increased the ratio of brain tryptophan to plasma free tryptophan if the rats were pretreated with the catecholamine reuptake inhibitor desipramine (Chaouloff et al. 1985a). Immobilization is likely to be more stressful than running because the animal cannot control the stress (Swenson and Vogel 1983; Weiss et al. 1981), while trained running is likely to be less stressful, because shock (and resultant stress) is avoided by running.

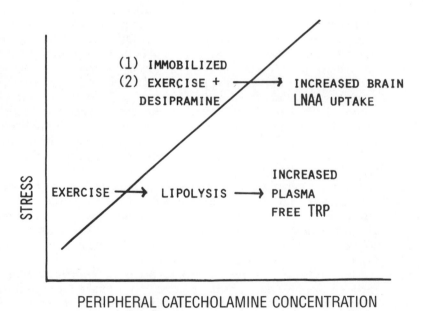

Figure 2-2. Suggested brain uptake mechanisms for tryptophan and other LNAAs activated by exercise and immobilization.

DISCUSSION

Tryptophan transport to the brain is under multiple influences, but specific influences predominate in different circumstances. Thus, food deprivation (Knott and Curzon 1972; Sarna et al. 1984) and exercise (Chaouloff et al. 1986b) increase brain tryptophan by increasing plasma free tryptophan; immobilization can increase it by altering LNAA transport kinetics (Kennett et al. 1986), and a high-carbohydrate meal increases it by decreasing plasma concentrations of competing LNAAs (Fernstrom and Wurtman 1971; Sarna et al. 1984). In other situations—aging, for example (Sarna et al. 1982), or hepatic failure (Bloxam and Curzon 1978)—combinations of these influences are involved.

Much of the work described in this chapter suggests that early ideas on the functional importance of physiologically relevant changes of brain transmitter precursor concentrations require some revision. It now seems likely that acute effects of feeding or of moderate periods of fasting ensure adequate brain stores of tryptophan (Curzon et al. 1972; Fernstrom and Wurtman 1971; Sarna et al. 1984) but do not normally have much effect on 5-HT concentrations at receptors, in particular at those that mediate appetite. Other recent findings on normal dietary changes of the ratio of plasma tryptophan to LNAAs are in agreement (Fernstrom 1987; Peters and Harper 1987). However, drug experiments show that pharmacological actions at 5-HT receptors can affect food intake (Hutson et al. 1988; Kennett and Curzon 1988; Kennett et al. 1987; Sugrue 1987). Therefore, it could be that brain tryptophan changes following dietary alterations, stress, or exercise are sufficient *in some individuals* to affect appetite or other behavioral components. Even laboratory rats of a single strain show remarkable individuality in the effects of tryptophan injection on their central tryptophan and 5-HT metabolism (Hutson et al. 1985), while considerable individual variation is seen in the magnitudes of the effects of serotonergic drugs on feeding in humans (Silverstone and Goodall 1986). It is also relevant that in their plasma amino acid studies on normal humans, Scriver et al. (1985) found striking homeostasis but large individual differences.

Thus, disorders of feeding could occur *in vulnerable subjects* through inappropriate effects of altered food intake on transmitter metabolism. For example, as administered tryptophan is reported to depress appetite in humans (Hrboticky et al. 1985), increases of brain tryptophan on food deprivation (Curzon et al. 1972; Knott and Curzon 1972) might conceivably be sufficiently large in susceptible individuals to cause anorexia. Similarly, increases of brain tryptophan in stress could have a role in some stress-induced anorexias. There is evidence that transport of tryptophan to the brain in stress can lead to altered brain function, as the competing LNAA valine opposed stress-induced rises of corticosterone (Joseph and Kennett 1983; Yehuda and Meyer 1984), an effect that was largely prevented by tryptophan (Joseph and Kennett 1983). These findings suggest that when brain tryptophan rises in stress, there could be an increase of 5-HT at hypothalamic 5-HT receptors (Haleem et al. 1989; Koenig et al. 1987) thought to mediate corticosterone secretion. Stress-induced increases of brain tryptophan may also augment morphine analgesia (Kelly and Franklin 1984), and it is, therefore, possible that exercise-induced increases (Chaouloff et al. 1985a, 1986a, 1986b) enhance mood-elevating effects of enkephalins released during exercise.

REFERENCES

Ahlborg G, Felig P, Hagenfeldt L, et al: Substrate turnover during prolonged exercise in man. J Clin Invest 53:1080–1090, 1974

Ashley DVM, Barclay DV, Chauffard FA, et al: Plasma amino acid responses in humans to evening meals of differing nutritional composition. Am J Clin Nutr 36:143–153, 1982

Ashley DVM, Liardon R, Leathwood PD: Breakfast meal composition influences plasma tryptophan to large neutral amino acid ratios of healthy lean young men. J Neural Transm 63:271–283, 1985

Banos G, Daniel PM, Moorhouse SR, et al: Inhibition of entry of some amino acids into the brain with observations on mental retardation in the aminoacidurias. Psychol Med 4:262–269, 1974

Belova TI, Jonsson G: Blood-brain barrier permeability and immobilization stress. Acta Physiol Scand 116:21–29, 1982

Bliss EL, Ailion J, Zwanziger J: Metabolism of norepinephrine, serotonin and dopamine in rat brain with stress. J Pharmacol Exp Ther 164:122–134, 1968

Bloxam DL, Curzon G: A study of proposed determinants of brain tryptophan concentration in rats after portacaval anastomosis or sham operation. J Neurochem 31:1255–1263, 1978

Carlsson A, Lindqvist M: Dependence of 5-HT and catecholamine synthesis on concentrations of precursor amino acids in rat brain. Naunyn Schmiedebergs Arch Pharmacol 303:157–164, 1978

Chaouloff F, Elghozi JL, Guezennec Y, et al: Effects of conditioned running on plasma, liver and brain tryptophan and on brain 5-hydroxytryptamine metabolism of the rat. Br J Pharmacol 86:33–41, 1985a

Chaouloff F, Laude D, Mignot E, et al: Tryptophan and serotonin turnover rate in the brain of genetically hyperammonaemic mice. Neurochemistry International 7:143–153, 1985b

Chaouloff F, Kennett GA, Serrurrier B, et al: Amino acid analysis demonstrates that increased plasma free tryptophan causes the increase of brain tryptophan during exercise in the rat. J Neurochem 46:1647–1650, 1986a

Chaouloff F, Laude D, Guezennec Y, et al: Motor activity increases tryptophan, 5-hydroxyindoleacetic acid and homovanillic acid in ventricular cerebrospinal fluid of the conscious rat. J Neurochem 46:1313–1316, 1986b

Cocchi D, di Giulio A, Groppetti A, et al: Hormonal inputs and brain tryptophan metabolism: the effect of growth hormone. Experientia 31:384–385, 1975

Curzon G: Methodological problems in the determination of total and free plasma tryptophan. J Neural Transm [Suppl] 15:221–226, 1979

Curzon G: Effects of food intake on brain transmitter amine precursors and amine synthesis, in Psychopharmacology and Food. Edited by Sandler M, Silverstone T. Oxford, UK, Oxford University Press, 1985, pp 59–70

Curzon G, Knott PJ: Fatty acids and the disposition of tryptophan, in Aromatic Amino Acids in the Brain (Ciba Foundation Symposium 22). New York, Elsevier, 1974, pp 217–229

Curzon G, Joseph MH, Knott PJ: Effects of immobilization and food deprivation on rat brain tryptophan metabolism. J Neurochem 19:1969–1974, 1972

Curzon G, Friedel J, Knott PJ: The effects of fatty acids on the binding of tryptophan to plasma protein. Nature 242:198–200, 1973

Daniel PM, Love ER, Pratt OE: Hypothyroidism and amino acid entry into brain and muscle. Lancet 2:872, 1975

Dunn AJ: Changes in plasma and brain tryptophan and brain serotonin and 5-hydroxyindoleacetic acid after foot shock stress. Life Sci 42:1847–1853, 1988

Enwonowu CO: Differential effect of total food withdrawal and dietary protein restriction on brain content of free histidine in the rat. Neurochem Res 12:483–487, 1987

Eriksson T, Carlsson A: Isoprenaline increases brain concentrations of administered L-DOPA and L-tryptophan in the rat. Psychopharmacology (Berlin) 77:98–100, 1982

Eriksson T, Granerus AK, Linde A, et al: "On-off" phenomenon in Parkinson's disease: relationship between dopa and other large neutral amino acids in plasma. Neurology 38:1245–1248, 1988

Felig P, Wahren J: Amino acid metabolism in exercising man. J Clin Invest 50:2703–2714, 1971

Fernstrom JD: Food induced changes in brain serotonin synthesis: is there a relationship to appetite for specific macronutrients? Appetite 8:163–182, 1987

Fernstrom JD, Wurtman RJ: Brain serotonin content: increase following ingestion of carbohydrate diet. Science 171:1023–1025, 1971

Fernstrom JD, Wurtman RJ: Brain serotonin content: physiological regulation by plasma neutral amino acids. Science 178:414–416, 1973

Fernstrom JD, Larin F, Wurtman RJ: Correlations between brain tryptophan and plasma neutral amino acid levels following food consumption in rats. Life Sci 13:517–524, 1973

Fernstrom JD, Fernstrom MH, Grubb PE: Twenty-four hour variations in rat blood and brain levels of the aromatic and branched-chain amino acids: chronic effects of dietary protein content. Metabolism 36:643–650, 1987

Gilman PK, Bartlett JR, Bridges PK, et al: Indolic substances in plasma, cerebrospinal fluid and frontal cortex of human subjects infused with saline or tryptophan. J Neurochem 37:410–417, 1981

Glaeser BS, Maher TJ, Wurtman RJ: Changes in brain levels of acidic, basic and neutral amino acids after consumption of single meals containing various proportions of protein. J Neurochem 41:1016–1021, 1983

Haleem DJ, Kennett GA, Whitton PS, et al: 8-OH-DPAT increases corticosterone but not other 5-HT$_{1A}$ receptor–dependent responses more in females. Eur J Pharmacol 164:435–443, 1989

Hawkins RA, Mans AM, Biebuyck JE: Amino acid supply to individual cerebral structures in awake and anaesthetized rats. Am J Physiol 242:E1–E11, 1982

Hrboticky N, Leiter LA, Anderson GH: Effects of L-tryptophan on short-term food intake in lean men. Nutr Res 5:595–607, 1985

Hutson PH, Knott PJ, Curzon G: Effect of isoprenaline infusion on the distribution of tryptophan, tyrosine and isoleucine, between brain and other tissues. Biochem Pharmacol 29:509–516, 1980

Hutson PH, Sarna GS, Kantamaneni BD, et al: Monitoring the effect of a tryptophan load on brain indole metabolism in freely moving rats by simultaneous cerebrospinal fluid sampling and brain dialysis. J Neurochem 44:1266–1273, 1985

Hutson PH, Dourish CT, Curzon G: Evidence that the hyperphagic response to 8-OH-DPAT is mediated by 5-HT$_{1A}$ receptors. Eur J Pharmacol 150:361–366, 1988

Joseph MH, Kennett GA: Corticosteroid response to stress depends upon increased tryptophan availability. Psychopharmacology (Berlin) 79:79–81, 1983

Kant GJ, Lenox RH, Bunnell BN, et al: Comparison of stress responses in male and female rats: pituitary cyclic AMP and plasma prolactic growth hormone and corticosterone. Psychoneuroendocrinology 8:421–428, 1983

Kelly SJ, Franklin KBJ: Evidence that stress augments morphine analgesia by increasing brain tryptophan. Neurosci Lett 44:305–310, 1984

Kennett GA, Curzon G: Evidence that hypophagia induced by mCPP and TFMPP requires 5-HT$_{1C}$ and 5-HT$_{1B}$ receptors; hypophagia induced by RU24969 only requires 5-HT$_{1B}$ receptors. Psychopharmacology (Berlin) 96:93–100, 1988

Kennett GA, Curzon G, Hunt A, et al: Immobilization decreases amino acid concentrations in plasma but maintains or increases them in brain. J Neurochem 46:208–212, 1986

Kennett GA, Dourish CT, Curzon G: 5-HT$_{1B}$ agonists induce anorexia at a postsynaptic site. Eur J Pharmacol 141:429–435, 1987

Knott PJ, Curzon G: Free tryptophan in plasma and brain tryptophan metabolism. Nature 239:452–453, 1972

Knott PJ, Joseph MH, Curzon G: Effects of food deprivation and immobilization on tryptophan and other amino acids in rat brain. J Neurochem 20:249–251, 1973

Knott PJ, Hutson PH, Curzon G: Fatty acid and tryptophan changes on disturbing groups of rats and caging them singly. Pharmacol Biochem Behav 7:245–252, 1977

Koenig JI, Gudelsky GA, Meltzer HY: Stimulation of corticosterone and β-endorphin secretion in the rat by selective 5-HT receptor subtype activation. Eur J Pharmacol 137:1–8, 1987

Langer P, Foldes O, Kvetnansky L, et al: Pituitary-thyroid function during acute immobilization stress in rats. Exp Clin Endocrinol 82:51–60, 1983

Mans AM, Biebuyck JF, Saunders SJ, et al: Tryptophan transport across the blood-brain barrier during acute hepatic failure. J Neurochem 33:409–418, 1979

Mans AM, Biebuyck JF, Davis DW, et al: Portacaval anastomosis: brain and plasma metabolite abnormalities and the effect of nutritional therapy. J Neurochem 43:697–705, 1984

Marcou M, Kennett GA, Curzon G: Enhancement of brain dopamine metabolism by tyrosine during immobilization: an in vivo study using repeated cerebrospinal fluid sampling in conscious rats. J Neurochem 48:1245–1251, 1987

Milakofsky L, Hare TA, Miller JM, et al: Rat plasma levels of amino acids and related compounds during stress. Life Sci 36:753–761, 1985

Milson JP, Morgan MY, Sherlock S: Factors affecting plasma amino acid concentrations in control subjects. Metabolism 28:313–319, 1979

Møller SE: Effect of various oral protein doses on plasma neutral amino acid levels. J Neural Transm 61:183–191, 1985

Mutch BJC, Banister EW: Ammonia metabolism in exercise and fatigue: a review. Med Sci Sports Exerc 15:41–50, 1983

Ohata M, Fredericks WR, Sundaram V, et al: Effect of immobilization stress on regional cerebral blood flow in the conscious rat. J Cereb Blood Flow Metab 1:187–194, 1981

Oldendorf WH, Szabo J: Amino acid assignment to one of three blood-brain amino acid carriers. Am J Physiol 230:94–98, 1976

Pardridge WM: Regulation of amino acid availability to the brain, in Nutrition and the Brain, Vol 1. Edited by Wurtman RJ, Wurtman JJ. New York, Raven Press, 1977, pp 141–204

Pardridge WM: Kinetics of competitive inhibition of neutral amino acid transport across the blood-brain barrier. J Neurochem 28:103–118, 1979a

Pardridge WM: Tryptophan transport through the blood-brain barrier: in vivo measurement of free and albumin-bound amino acid. Life Sci 25:1519–1528, 1979b

Pardridge WM: Potential effects of the dipeptide sweetener aspartame on the brain, in Nutrition and the Brain, Vol 7. Edited by Wurtman RJ, Wurtman JJ. New York, Raven Press, 1986, pp 199–241

Perez-Cruet J, Chase TN, Murphy DL: Dietary regulation of brain tryptophan metabolism by plasma ratio of free tryptophan and neutral amino acids in humans. Nature 248:693–695, 1974

Peters JC, Harper AE: Acute effects of dietary proteins on food intake, tissue amino acids and brain serotonin. Am J Physiol 252:R902–R914, 1987

Rigotti P, Jonung T, Peters JC, et al: Methionine sulfoximine prevents the accumulation of large neutral amino acids in brain of portacaval shunted rats. J Neurochem 44:929–933, 1985

Sarna GS, Tricklebank MD, Kantamaneni BD, et al: Effect of age on variables influencing the supply of tryptophan to the brain. J Neurochem 39:1283–1290, 1982

Sarna GS, Kantamaneni BD, Curzon G: Variables influencing the effect of a meal on brain tryptophan. J Neurochem 44:1575–1580, 1984

Scriver CR, Gregory DM, Sovetts D, et al: Normal plasma free amino acid values in adults: the influence of some common physiological variables. Metabolism 34:868–873, 1985

Shamoon H, Jacob R, Sherwin RS: Epinephrine induced hypoaminoacidemia in normal and diabetic subjects: effects of blockade. Diabetes 11:875–881, 1980

Silverstone T, Goodall E: Serotonergic mechanisms in human feeding: the pharmacological evidence. Appetite 7 (suppl): 85–97, 1986

Sugrue MF: Neuropharmacology of drugs affecting food intake. Pharmacol Ther 32:145–182, 1987

Sved AF: Precursor control of the function of monoaminergic neurons, in Nutrition and the Brain, Vol 6. Edited by Wurtman RJ, Wurtman JJ. New York, Raven Press, 1983, pp 224–263

Swenson RM, Vogel WH: Plasma catecholamine and corticosterone as well as brain catecholamine changes during coping in rats exposed to stressful foot shock. Pharmacol Biochem Behav 18:689–693, 1983

Tang LC, Cotzias GC: Modifications of the actions of some neuroactive drugs by growth hormone. Arch Neurol 33:131–134, 1976

Weiss JM, Goodman PA, Losito BC, et al: Behavioural depression produced by an uncontrollable stressor: relationship to norepinephrine, dopamine and serotonin levels in various regions of rat brain. Brain Res Rev 3:167–205, 1981

Yehuda R, Meyer JS: A role for serotonin in the hypothalamic-pituitary adrenal response to insulin stress. Neurochemistry 38:25–32, 1984

Yokohoshi H, Wurtman RJ: Meal composition and plasma amino acid ratios: effect of various proteins or carbohydrates and of various protein concentrations. Metabolism 35:837–842, 1986

Chapter 3

Factors Influencing the Therapeutic Effect of Tryptophan in Affective Disorders, Sleep, Aggression, and Pain

Simon N. Young, Ph.D.

Chapter 3

Factors Influencing the Therapeutic Effect of Tryptophan in Affective Disorders, Sleep, Aggression, and Pain

The use of tryptophan has been studied for over 30 years. The attraction of this topic is that it may have both theoretical and practical significance: theoretical because tryptophan is relatively specific in its effects on serotonin and practical because tryptophan has relatively few side effects and can have a useful therapeutic effect in some circumstances.[1]

---

[1] After the completion of this article there were reports of serious side effects associated with the use of tryptophan in the United States. The cluster of adverse effects has been named the eosinophilia-myalgia syndrome (EMS). EMS is characterized by intense muscle and joint pain accompanied by fatigue and eosinophilia. Other symptoms include fever and skin rashes. Onset is relatively abrupt and usually follows ingestion of tryptophan for weeks or months. At the time of writing (April 1990) over 1,000 cases and more than 20 deaths associated with EMS have been reported in the United States. In Canada nine cases have been reported. All nine bought their supplies of tryptophan over the counter in the United States. No cases have been reported associated with the brand of tryptophan sold in Canada as a prescription drug for the treatment of depression. In Britain three cases of EMS have been reported. In all three cases the tryptophan was obtained in the United States. No cases have been reported associated with the brands of tryptophan sold in Britain as a prescription drug for the treatment of depression. These data point to the conclusion that EMS is associated with an impurity in one or more of the brands of tryptophan sold over the counter in the United States. However, at the time of writing this idea still remains a working hypothesis rather than a definite conclusion.

In 1958 Lauer et al. published the results of a study in which they gave L-tryptophan (0.2 mg/kg/day for 6 weeks) to seven schizophrenic patients who were also receiving the monoamine oxidase inhibitor (MAOI) iproniazid. They argued that "if the degradation of serotonin could be impeded by a monoamine oxidase inhibitor, and its synthesis simultaneously accelerated by giving the precursor from which it is formed, effects might be obtained which were more pronounced." That tryptophan administration does actually increase serotonin synthesis was shown 3 years later in rats (Hess and Doepfner 1961) and 12 years later in humans (Eccleston et al. 1970). The behavioral observations of Lauer et al. (1958) that "the patients exhibited an increase in energy level and motor activity and improvement in the ability to accept interpersonal relationships, and displayed more affect" have been the starting point for many studies over the subsequent 30 years of the effect of tryptophan on mood.

Another important milestone is the study of Olson et al. (1960). As part of a clinical biochemical study, Olson and colleagues gave 10 g of DL-tryptophan to normal subjects. They noted various side effects, "mainly euphoria and lightheadedness, but including dizziness, headache and rarely nausea, which lasted for one or two hours." This observation was followed up by Smith and Prockop (1962), who gave L-tryptophan (90 mg/kg) to normal subjects specifically to study effects on central nervous system function. The main effects were alternating euphoria and drowsiness. These observations led to the largest body of studies on the clinical effects of tryptophan, those on arousal and on sleep.

The studies on the action of tryptophan in affective disorders and insomnia resulted from clinical observation. However, animal studies have shown that serotonin can influence a wide variety of behaviors, including aggression, pain perception, and food intake. These observations have led to studies on the action of tryptophan in pathological aggression, pain, and eating disorders. However, these topics have attracted less interest, and only studies on aggression and pain look sufficiently promising to be discussed here.

The purpose of this chapter is to discuss the role of tryptophan in the treatment of affective disorders, pathological aggression, insomnia, and pain. However, an understanding of the biochemistry and physiology of tryptophan is important to optimize its clinical use and can in some cases throw light on apparent discrepancies in the clinical literature. Therefore, this chapter starts with a discussion of tryptophan in the diet, of its peripheral metabolism, and of its effect on serotonin synthesis and function, before assessing the clinical studies.

Young (1986) has published a more detailed review of the clinical psychopharmacology of tryptophan.

TRYPTOPHAN AND DIET

To those unused to therapy with amino acids, the doses of tryptophan used may seem massive. However, even the normal daily dietary intake of tryptophan is much larger than the daily dose of many drugs. An adult male needs 0.25 g tryptophan to maintain nitrogen balance (Altman and Dittmer 1968), while a normal diet contains 1 to 1.5 g tryptophan per day (Cole et al. 1980; Murphy et al. 1974). The highest doses needed when tryptophan is given chronically are around 6 to 8 g per day, which does not exceed eight times the normal daily dietary intake.

One question that often arises is whether diets high in tryptophan could duplicate the effects of ingesting tryptophan in purified form. For example, is the traditional belief in the hypnotic effect of a glass of warm milk taken before bedtime due to the tryptophan in the proteins of milk? The answer to this question is a clearcut no. In the rat, protein-containing meals tend to lower brain tryptophan and serotonin (Fernstrom and Faller 1978; Wurtman et al. 1981) because all large neutral amino acids (LNAAs), including tryptophan, are transported across the blood-brain barrier by the same transport system. The various amino acids compete for the carrier (Oldendorf and Szabo 1976). Therefore, the brain level of tryptophan will vary in proportion to the plasma ratio of tryptophan to the sum of the other amino acids that compete for the transport system. When a protein meal is ingested, plasma tryptophan will rise. However, since tryptophan is the least abundant amino acid in most proteins, the plasma levels of the other LNAAs will rise even more, creating greater competition for the blood-brain barrier transport system, so brain tryptophan and serotonin will decline. No proteins are high enough in tryptophan, relative to the other LNAAs, to raise brain tryptophan. In humans the same sort of mechanism operates as in rats, but the changes in the plasma tryptophan ratio tend to be smaller. In one study ingestion of a protein-containing meal by patients caused a small decline in cerebrospinal fluid (CSF) tryptophan and 5-hydroxyindoleacetic acid (5-HIAA) (Perez-Cruet et al. 1974), but in another study a protein meal did not influence the plasma ratio enough to alter CSF tryptophan or 5-HIAA (Teff et al. 1989).

Although the effects of tryptophan supplementation cannot be duplicated by a diet high in tryptophan, dietary factors may influence the rise of brain tryptophan after ingestion of purified tryptophan. If tryptophan is taken with a balanced meal containing protein, the

LNAAs in the protein may blunt the initial rise in brain tryptophan. As discussed below, the initial rise in brain tryptophan after tryptophan ingestion is probably large enough to saturate tryptophan hydroxylase and maximize the rate of brain serotonin, when tryptophan is given in single doses of 2 to 3 g. As this is the type of single dose given tid when tryptophan is given for the treatment of affective disorders, taking it with meals will not block its therapeutic effect and may diminish side effects.

Tryptophan is effective as a hypnotic at doses as low as 1 g. At this dosage level, taking it with a protein-containing meal may block its activity. On the other hand, taking it with pure carbohydrate may be beneficial because carbohydrate causes release of insulin, which stimulates the uptake of the branched-chain amino acids, leucine, isoleucine, and valine into muscle. Their plasma levels fall, there is less competition for the entry of tryptophan into brain, and brain tryptophan and serotonin will tend to increase (Fernstrom and Wurtman 1971; Wurtman et al. 1981). Carbohydrate meals cannot duplicate the effects of tryptophan administration because the increase of brain tryptophan and serotonin is smaller and of shorter duration after carbohydrate than after tryptophan. However, taking tryptophan as a hypnotic with pure carbohydrate may enable smaller doses of tryptophan to be used. As tryptophan is metabolized relatively quickly, low doses have the advantage that they will be removed from the body relatively quickly after they have had their hypnotic effect. In the rat, as little as 4% protein in a meal is sufficient to block the effect of carbohydrate on brain tryptophan (Yokogoshi and Wurtman 1986). Thus, only relatively pure carbohydrate intake will potentiate the uptake of tryptophan into brain.

Because tryptophan is a dietary component, it is often regarded as a "natural" or "dietary" treatment. However, as discussed above, the effects on brain serotonin of dietary tryptophan and of purified tryptophan capsules or tablets are fundamentally different. When tryptophan supplements are given, it is for the purpose of raising brain serotonin rather than for correcting any dietary deficiency. In these circumstances "drug" would seem to be a more apt description of tryptophan than "dietary supplement." However, this issue is treated differently in different countries. In the United States tryptophan is freely available as a dietary supplement in health food stores, drug stores, and even supermarkets. In Britain and Canada it is available only as a prescription drug, under the brand names Optimax and Pacitron in Britain and Tryptan in Canada. In Canada tryptophan was freely available as a dietary supplement until 1985. The change in its status prompted a legal challenge, but the court ruled in favor of the government.

PERIPHERAL METABOLISM OF TRYPTOPHAN

The main catabolic pathway of tryptophan is initiated by tryptophan-2,3-dioxygenase (tryptophan pyrrolase) in the liver and indoleamine-2,3-dioxygenase (indolamine pyrrolase), which is found in a variety of tissues (Hayaishi 1985). The pathway started by these enzymes leads through kynurenine to CO_2. A variety of other important compounds are also formed, including quinolinic acid, kynurenic acid, 3-hydroxyanthranillic acid, and niacin. The activity of tryptophan pyrrolase increases markedly upon tryptophan loading, and this limits the rise in brain tryptophan and serotonin (Young and Oravec 1979). In patients receiving a 50 mg/kg dose of tryptophan, an amount often given as a single dose in clinical studies, CSF tryptophan and 5-HIAA remain elevated for only 12 hours (Eccleston et al. 1970). Increasing the dose of tryptophan speeds its catabolism even more and shortens the plasma half-life of tryptophan (Green et al. 1980). As a result, while 50 mg/kg tryptophan causes a peak 9-fold increase in plasma tryptophan, doubling the dose to 100 mg/kg increases the peak value only 12-fold above baseline (Yuwiler et al. 1981). However, higher doses of tryptophan are capable of prolonging the rise of CSF 5-HIAA somewhat (Young and Gauthier 1981).

Inhibition of the metabolism of tryptophan would enable lower and less frequent doses to be given. However, allopurinol and nicotinamide, two compounds that can inhibit tryptophan catabolism in the rat, failed to influence human plasma tryptophan levels after a tryptophan load when given in clinically acceptable doses (Green et al. 1980; Møller 1981; Møller and Kirk 1978). In the rat, high doses of vitamin B6 can inhibit tryptophan degradation and increase brain tryptophan (Bender and Totoe 1984). The effect of pyridoxine on plasma tryptophan in humans has not been studied, but pyridoxine should be given with tryptophan anyway, for another reason. Several of the enzymes on the catabolic pathway require pyridoxal phosphate as a cofactor. Chronic tryptophan treatment may increase requirements for vitamin B6, as supplements of pyridoxine were found to attenuate the increase in urinary tryptophan metabolites that occurred when normal subjects were given tryptophan for a week (Green and Aronson 1980). As some of the tryptophan metabolites may have adverse effects (see section on toxicity below), supplements of vitamin B6 should be given with tryptophan.

In some of the earlier clinical studies, when L-tryptophan was not freely available, the DL isomer was used. Although the rat can convert the D isomer to the L form, humans cannot (Hankes et al. 1972). Therefore, when the DL isomer is used, the effective dose is half the actual dose. In this chapter tryptophan always refers to L-tryptophan unless otherwise specified.

TRYPTOPHAN, SEROTONIN SYNTHESIS, AND SEROTONIN FUNCTION

The rationale for the therapeutic use of tryptophan is that the amino acid will increase serotonin synthesis and function. Increases in serotonin synthesis undoubtedly occur, but they are limited in their extent. In neurological patients, a 3-g load of tryptophan caused a doubling of CSF 5-HIAA, while a 6-g load caused no further increase (Young and Gauthier 1981), suggesting that in humans, as in rats, tryptophan hydroxylase is normally about half saturated with tryptophan. Therefore, tryptophan loading can do no more than double the rate of serotonin synthesis. Whether such an increase in serotonin synthesis will also enhance release of serotonin from neurons (i.e., serotonin function) is more problematic.

Under normal circumstances the amount of serotonin released over any period of time will depend on the amount of serotonin released each time a neuron fires and on the rate of firing of the neurons. The cell bodies of serotonergic neurons are located in the raphe nuclei in the brain stem. Single unit recording of rat raphe nuclei cells reveals a slow rhythmic discharge rate (Gallagher and Aghajanian 1976). When rats are given tryptophan, there is a dose-dependent decrease in the firing rate (Trulson and Jacobs 1976), which is dependent on increased serotonin levels in the raphe nuclei (Gallagher and Aghajanian 1976). Obviously a decrease in the firing rate of serotonergic neurons will tend to counteract any effect of increased serotonin levels that might promote increased release of the neurotransmitter. This fact might suggest that any effect of tryptophan on serotonin release will be minimal unless other factors increase the firing rate of serotonin neurons, or serotonin release occurs independent of neuronal firing. In the cat, raphe cell firing rates increase with the level of behavioral arousal of the animal (Trulson and Jacobs 1979), suggesting that tryptophan might be more effective when given to patients who are highly aroused, a possibility that is discussed further in later sections of this chapter. The other possible mechanism to increase serotonin release is to cause release of the neurotransmitter independent of neuronal firing. There is behavioral evidence that this phenomenon occurs in rats when serotonin synthesis is increased by tryptophan loading at the same time as its degradation is inhibited by an MAOI. Under these circumstances serotonin is thought to "spill over" onto postsynaptic receptors (Grahame-Smith 1971). In keeping with this idea, when rats are given an MAOI their CSF serotonin, which is extracellular and must have been released from neurons, increases threefold. However, the combination of an MAOI plus tryptophan raises CSF serotonin more than 20-fold (Anderson et al. 1987). The clinical use of tryptophan plus an MAOI is discussed in a later section.

Although the most likely mechanism for the behavioral effects of tryptophan is an increase in serotonin release from neurons, other mechanisms should also be considered. Tryptophan might inhibit brain uptake of phenylalanine, tyrosine, or histidine and therefore decrease brain catecholamines or histidine. However, in psychiatric patients a single 5 g tryptophan dose failed to lower CSF tyrosine (van Praag et al. 1973), so tryptophan is unlikely to decrease catecholamines or histamine. Tryptophan is the precursor of a variety of psychoactive compounds other than serotonin, such as tryptamine, melatonin, quinolinic acid, and kynurenic acid. However, tryptophan has not been shown to influence these compounds enough to alter any aspect of brain function. Although these alternative mechanisms should be kept in mind and could perhaps contribute to some of tryptophan's actions, at the moment the most plausible working hypothesis is that tryptophan acts through alterations in serotonin release.

TOXICITY AND SIDE EFFECTS OF TRYPTOPHAN

The side effects of tryptophan are mild. The ones most commonly observed are nausea, dizziness, headache, and drowsiness (Young 1986). In many of the clinical studies no side effects were reported, even when the doses used were as high as 9.6 g/day (Murphy et al. 1974) or even 20 g/day (Gillin et al. 1976). In double-blind, placebo-controlled studies, which included a comparison of the side effects of tryptophan and placebo, no significant differences were seen (Chouinard et al. 1985; Thomson et al. 1982). Although tryptophan produces only minor side effects when given by itself, in combination with an MAOI symptoms can be more severe. The most common effects of this combination are dizziness, nausea, and headache (Ayuso Gutierrez and Lopez-Ibor Alino 1971; Coppen et al. 1963; Glassman and Platman 1969), but other effects, including ethanol-like intoxication, drowsiness, hyperreflexia, and clonus, are also seen (Oates and Sjoerdsma 1960).

The main concern about toxicity of tryptophan is the possible carcinogenic action of the tryptophan metabolite 3-hydroxyanthranillic acid. This compound causes bladder cancer when it is planted in pellet form in the bladder of the rat (Bryan 1971). However, in a large study carried out by the National Cancer Institute, tryptophan was not found to produce cancer in either rats or mice (National Cancer Institute 1978). Therefore cancer should only be of concern in patients who have a source of physical irritation in their bladder, as in the experiment with implantation of pellets of 3-hydroxyanthranillic acid. The tryptophan metabolite xanthurenic acid has a mild diabetogenic action in animals, possibly because of its ability to bind insulin (Hattori et al. 1984; Ikeda and Kotake 1984), suggesting that patients with a family history of diabetes should be followed closely. Supplementation of tryptophan in the diet of pregnant hamsters caused significant reductions in embryo and neonate survival and in neonatal weight of the pups (Meier and Wilson 1983). Thus, the idea of using a "natural" compound such as tryptophan in pregnant women is definitely contraindicated.

A recent report describes ultrastructural changes in the liver following the oral administration of 250 mg/kg of tryptophan per day for 3 days to rats. This treatment produced enlarged hepatic sinusoids and vacuolated cells, many of which contained lipid (Trulson and Sampson 1986). The relevance of these findings to humans is not clear. In Britain, where tryptophan has been on the market as an antidepressant for many years, very few adverse effects have been reported. As Hartmann (1987) has pointed out, in the National Cancer Institute (1978) study, very large doses of tryptophan were given to rats for 78 weeks; yet, no significant gross pathology of the liver was seen and the lifespan of the rats was not decreased. Thus, the report of tryptophan-induced liver damage in the rat should not preclude its careful use in humans.

More detailed reviews on the toxicity of tryptophan are available (Sourkes 1983; Young 1986).

THE ANTIDEPRESSANT ACTION OF TRYPTOPHAN

Tryptophan and Monoamine Oxidase Inhibitors

The first test of the antidepressant action of tryptophan was published by Coppen et al. in 1963. They gave 10 to 17 g of DL-tryptophan or placebo under double-blind conditions to depressed patients during their second week of treatment with the MAOI tranylcypromine. Improvement in the tryptophan group was significantly more rapid than in the placebo group. Three other placebo-controlled studies have confirmed that tryptophan is capable of potentiating the antidepressant action of MAOIs (Ayuso Gutierrez and Lopez-Ibor Alino 1971; Glassman and Platman 1969; Pare 1963). This finding is consistent with the fact, discussed above, that the combination of tryptophan and an MAOI causes a large release of serotonin that is independent of neuronal firing. Because tryptophan can potentiate the side effects of MAOIs as well as their therapeutic action, this combination of drugs is usually used only in treatment-resistant patients.

Tryptophan and Other Antidepressant Treatments

The addition of tryptophan to amine uptake inhibitors does not produce the same clear effect that is seen when it is added to MAOIs. Tryptophan was not found to potentiate the action of imipramine or amitriptyline (Pare 1963), clomipramine or desipramine (Shaw et al. 1975), amitriptyline (Chouinard et al. 1979b), or zimelidine (Walinder et al. 1981). However, it produced a nonsignificant potentiation of amitriptyline in one study (Lopez-Ibor Alino et al. 1973), and in another the combination of clomipramine plus tryptophan produced a more rapid and greater improvement in depression and anxiety than clomipramine and placebo (Walinder et al. 1976). The largest and longest study on the combination was carried out by Thomson et al. (1982) on mildly to moderately depressed outpatients, unlike the other studies which were mostly on severely depressed inpatients. The Hamilton Depression Rating Scale scores showed no differences for amitriptyline, tryptophan, and their combination, but the scale item "depressed mood" showed significantly greater improvement for the combination than for either treatment alone.

The advantages of adding tryptophan to amine uptake inhibitors are relatively small. However, side effects of uptake inhibitors are not enhanced by tryptophan. Indeed, Thomson et al. (1982) found that the increase in heart rate seen with amitriptyline was much less marked in the combined treatment group.

The large increase in serotonin release seen with tryptophan and an MAOI would not be expected to occur with the combination of tryptophan and an amine uptake inhibitor. Thus the clinical data are consistent with what is known about the pharmacological action of these drugs on serotonin.

The addition of tryptophan to electroconvulsive therapy (ECT) has been tested in two studies. In one, tryptophan was no better than placebo when infused intravenously before each ECT (Kirkegaard et al. 1978). In the other, oral tryptophan (6 g/day), given from 1 day before the first ECT until 4 days after the last ECT, was compared with placebo (d'Elia et al. 1977). Tryptophan caused a significant potentiation of the antiretardation effect of ECT, but this effect was considered of little importance clinically. The combination of tryptophan with lithium was considered effective in two studies (Honore et al. 1982; Worrall et al. 1979), but the absence of suitable control groups makes it impossible to say whether tryptophan was increasing the antidepressant effect of lithium. The same can be said of a study in which seven treatment-resistant patients, who had not responded to lengthy trials of tricyclics and MAOIs, were successfully treated with the combination of clomipramine, lithium, and tryptophan (Hale et al. 1987).

Tryptophan by Itself in Depression

The confusing picture from the large number of studies of tryptophan given without other antidepressant treatments has been reviewed (Baldessarini 1984; Gelenberg et al. 1982; Young 1986). Interpretation of this literature is difficult because different studies use different designs, dosages, and types of patients and reach different conclusions.

Most of the studies with no control group found little therapeutic effect (Bowers 1970; Gayford et al. 1973; Mendels et al. 1975; Worrall et al. 1979). Two found no response in unipolar patients but improvement in half of a small number of bipolar patients (Farkas et al. 1976; Murphy et al. 1974). However, three studies achieved a good response in three-quarters of the patients given tryptophan (Broadhurst 1970; Chouinard et al. 1983; Shopsin 1978). Overall, these studies indicate that tryptophan has little effect, as the positive studies can be explained more easily, by placebo or study effects, than can the negative studies.

Many studies have found tryptophan not significantly different from standard antidepressant treatments such as imipramine (Broadhurst 1970; Chouinard et al. 1979b; Coppen et al. 1972; Kline and Shah 1973; Lindberg et al. 1979; Rao and Broadhurst 1976), amitriptyline (Herrington et al. 1976), and mianserin (Bennie 1982). However, the sample sizes in these studies would not necessarily have been large enough to demonstrate an antidepressant effect of imipramine versus placebo. In a more demanding comparison against ECT, tryptophan was inferior in two studies (Carroll et al. 1970; Herrington et al. 1974), and of equal efficacy in one study (Coppen et al. 1967), while 3 g of tryptophan per day was found to be superior to unilateral ECT administered twice weekly in one study (MacSweeney 1975). There does not seem to be any way of reconciling this discrepancy.

In three small studies with placebo controls there was no evidence for any therapeutic effect of tryptophan (Chouinard et al. 1983;

Mendels et al. 1975; Murphy et al. 1974). In larger placebo-controlled studies on special types of depressed patients tryptophan was found to have no effect on patients with maternity blues (Harris 1980) or on geriatric patients with mild to moderate depression of mixed etiology (endogenous, reactive, and secondary to organic disease) (Cooper and Datta 1980).

Nearly all the studies described above were performed on severely depressed inpatients. Because of the wide variety of results achieved, any conclusion must be tentative. In the majority of severely depressed patients tryptophan does not seem to be an effective antidepressant. However, in a minority of patients, possibly bipolar, it may have a useful effect. Although tryptophan has limited efficacy in severely depressed patients, it may have a useful therapeutic effect in less depressed patients. The longest (12 weeks) and largest (27 to 31 patients per group) study on tryptophan was a double-blind comparison of placebo, tryptophan (3 g per day), amitriptyline, and tryptophan plus amitriptyline in mildly to moderately depressed outpatients treated by general practitioners (Thomson et al. 1982). Tryptophan was significantly better than placebo and equivalent to amitriptyline in therapeutic efficacy. Moreover, tryptophan had significantly fewer side effects than amitriptyline. Because of this study's size, its results are more convincing than those of the other studies. Overall, this suggests that tryptophan is a useful antidepressant in mildly or moderately depressed patients but not in patients with severe depression.

As discussed above, serotonin neurons fire at faster rates when animals are at a high level of arousal. The effects of tryptophan on serotonin release should be greater when serotonin neurons are firing at a faster rate. In severe depression, psychomotor retardation is often seen. In patients with psychomotor retardation, serotonin neurons are presumably firing at a slow rate. Thus, the effect of tryptophan on serotonin release may be small, the result being little therapeutic effect. In mildly or moderately depressed outpatients the degree of psychomotor retardation will be less than in severely depressed patients. Thus, their serotonin neurons may be firing at a faster rate, and this may explain the therapeutic effect of tryptophan in these patients.

TRYPTOPHAN IN MANIA

There are several reasons for testing tryptophan in mania. Increased serotonin action decreases the response of experimental animals to a variety of stimuli, and enhanced responses to stimuli are a characteristic of manic patients. As discussed in the section on sleep below, tryptophan can have a sedative action, while low serotonin has been suggested as one factor in the etiology of mania (Prange et al. 1974). The elevated levels of arousal in mania should ensure that tryptophan enhances serotonin release.

Murphy et al. (1974) gave 10 manic patients an average dose of 9.6 g tryptophan per day. Of seven who improved, three subsequently relapsed on placebo. Prange et al. (1974) compared tryptophan with moderate doses of chlorpromazine and found a slight superiority of tryptophan in all respects. In a comparison with placebo, no effect of tryptophan was seen, but this is not surprising as there were only five patients per group and the patients also received nitrazepam and chlorpromazine prn (Chambers and Naylor 1978). In a modified crossover design, Chouinard et al. (1985) achieved a better therapeutic effect with tryptophan than with placebo. Overall these studies suggest a clinically useful effect of tryptophan in acute mania, but the effect is certainly not as strong as that of high-dose neuroleptics. The most useful role for tryptophan may be in conjunction with lithium. Brewerton and Reus (1983) gave tryptophan or placebo to a group of manic and schizoaffective patients receiving lithium and saw a significant improvement in patients on tryptophan relative to those on placebo. This is consistent with electrophysiological evidence that tryptophan enhances central serotonin transmission in lithium-treated animals (Sangdee and Franz 1980).

Three single-case studies of bipolar patients indicate a prophylactic action of tryptophan either in conjunction with lithium (Chouinard et al. 1979a) or by itself (Beitman and Dunner 1982; Hertz and Sulman 1968). These interesting leads need to be followed up with controlled studies.

TRYPTOPHAN AND AGGRESSION

Studies on 5-HIAA in CSF indicate low serotonin in patients who are aggressive toward others and in the autoaggression of suicide (Åsberg et al. 1987). This provides a rationale for the treatment of aggression or suicidal ideation with tryptophan. Tryptophan has not yet been tested on suicidal ideation, and only a limited amount of data is available on the therapeutic action of tryptophan in pathological aggression. Morand et al. (1983), using a crossover design, gave tryptophan and placebo to 12 male aggressive schizophrenic patients (mainly murderers) who had not responded to neuroleptics. Tryptophan caused a significant decrease in incidents as measured by a ward checklist that recorded physical aggression, verbal abuse, and other uncontrolled behaviors.

The combination of tryptophan and the serotonin uptake inhibitor trazodone has been tested in several aggressive patients who had not responded to other drugs. Good response was seen in six of seven patients who exhibited aggression in association with senile dementia (Greenwald et al. 1986; Watson and Phillips 1986; Wilcock et al. 1987) and in one patient who was mentally retarded (O'Neil et al. 1986).

More work is needed on the effect of tryptophan in aggression. Patients with impulsive aggression might be expected to respond best both because impulsivity is associated with low CSF 5-HIAA and because the arousal associated with impulsivity may enhance tryptophan-mediated release of serotonin.

TRYPTOPHAN AND SLEEP

Smith and Prockop (1962) found that tryptophan produced alternating euphoria and drowsiness in normal subjects, and this observation has been followed up by more than 40 studies looking at the effect of tryptophan on sleep. Detailed reviews of these studies are available (Hartmann and Greenwald 1984; Schneider-Helmert and Spinweber 1986).

Tryptophan can in some circumstances alter sleep stages. For example, in an early study, Wyatt et al. (1970) gave 7.5 g tryptophan to normal subjects and found a decrease in rapid eye movement (REM) sleep and an increase in non-REM sleep. Other studies have found no effect on sleep stages, and this certainly seems to be true at lower doses; for example, Hartmann et al. (1974) found no alteration in sleep stages with doses below 5 g, but a decrease in desynchronized sleep and an increase in slow-wave sleep with higher doses.

The most commonly observed effect of tryptophan is a decrease in sleep latency. This effect is not always seen, but Hartmann and Greenwald (1984) and Schneider-Helmert and Spinweber (1986) have identified some of the factors that influence its appearance. Dose is certainly important. While doses above 1 g seem to decrease sleep latency with equal efficacy (Hartmann et al. 1974), doses of 0.25 and 0.5 g showed only a nonsignificant trend to decreased latency (Hartmann and Spinweber 1979). Time of administration is another important variable. When subjects who received 4 g tryptophan filled out the Stanford Sleepiness Scale every 15 minutes, a significant increase in sleepiness started only at 45 minutes after tryptophan administration (Hartmann et al. 1976). The type of subject also influences the response to tryptophan. Tryptophan, given at an adequate dose, decreases sleep latency in subjects with mild insomnia or in subjects with a long sleep latency who do not complain of insomnia. Effects are far less likely to be seen in normal subjects (where there is little scope for decreasing sleep latency) and in subjects with chronic or severe insomnia. In subjects with chronic or severe insomnia, tryptophan, given acutely, is significantly less effective than standard hypnotics (Hartmann et al. 1983; Linnoila et al. 1980).

Evidence is slowly accumulating that, in patients with severe insomnia, the efficacy of tryptophan may develop slowly over a matter of days, even though it is given only in a single dose at bedtime. Thus, when tryptophan is given for several nights followed by placebo for several nights, a decrease in sleep latency is often seen in the latter part of the tryptophan treatment period, as well as in the post-treatment placebo period (relative to the pretryptophan baseline) (Gnirss et al. 1978; Hartmann et al. 1983; Schneider-Helmert 1981; Spinweber 1986; Wyatt et al. 1970). This effect has not yet been firmly established by a study comparing the effect of tryptophan followed by placebo with that of placebo followed by placebo but nonetheless seems to be a promising approach. Schneider-Helmert and Spinweber (1986) have suggested that interval therapy should always be used when tryptophan is given as a hypnotic, with several days of tryptophan treatment alternating with several days off tryptophan. If such an approach is effective it will be important because of the lack of side effects of tryptophan when given as a hypnotic. Thus, Spinweber (1986) found that tryptophan did not alter sleep stages, impair performance, elevate the threshold for arousal from sleep, or alter brain electrical activity during sleep.

TRYPTOPHAN AND PAIN

A large body of data from experiments on animals suggests that serotonin in the spinal cord can decrease nociceptive afferents and is involved in the analgesic action of morphine. Thus, it is unfortunate that only a small number of studies have looked at the effect of tryptophan on pain and that the majority of clinical studies in this area involved only a small number of patients and inadequate controls.

Tests of tryptophan in experimental pain have given inconsistent results. Seltzer et al. (1982) gave tryptophan or placebo to 30 normal subjects. Perception threshold after electrical stimulation of dental pulp was unchanged by tryptophan, but it increased pain tolerance. Using a signal detection method, Lieberman et al. (1983) found that tryptophan reduced the discriminability of thermal pain stimuli. However, 14 days of tryptophan treatment failed to alter radiant heat pain thresholds in 20 female student volunteers (Mitchell et al. 1987), while tryptophan reduced experimentally induced ischemic pain in only 3 out of 11 subjects (Nurmikko et al. 1984).

Results with clinical pain also have been inconsistent. In a double-blind study Seltzer et al. (1983) gave tryptophan or placebo for 4 weeks to 30 patients with chronic maxillofacial pain. There was a greater reduction in reported clinical pain and a greater increase in pain tolerance in the tryptophan group than in the placebo group. Moldofsky and Lue (1980) gave tryptophan or chlorpromazine to patients with fibrositis syndrome. Chlorpromazine, but not tryptophan, was associated with increased slow-wave sleep and amelioration of pain and mood. As tryptophan was given as a single 5-g dose at bedtime, the lack of effect on pain is not surprising. However, Sternbach et al. (1976) found that chlorimipramine, but not tryptophan, was effective in decreasing pain in patients with disc disease. Shpeen et al. (1984) found that 0.5 g of tryptophan given every 6 hours was significantly better than placebo in relieving pain 24 hours after nonsurgical endodontic therapy.

In three small open studies tryptophan was found effective in reversing the decline in effectiveness over time of other analgesic treatments. In four out of five patients tryptophan, but not placebo, given at a dose of 3 g per day for 2 months reversed the tolerance that occurred when severe intractable pain was treated by electrical stimulation of periaqueductal and periventricular gray matter (Hosobuchi 1978). In three of these patients tryptophan was shown to restore the release of beta-endorphin immunoreactivity into the ventricular CSF, which had occurred initially on electrical stimulation but had stopped when tolerance developed to the analgesic effect of electrical stimulation (Hosobuchi et al. 1980b). Tryptophan also reversed tolerance to opiates in five patients who had been given opiates chronically to treat low back and leg pain (Hosobuchi et al. 1980a). Tryptophan was given to five rhizotomy and cordotomy patients whose pain had resumed and whose sensory deficits had diminished. Their sensory deficits for both touch and pinprick reexpanded to the maximum extent initially recorded after surgery (King 1980).

There is no simple way of reconciling the results of these different studies. Pain is a complex phenomenon and different types of pain may have different mediating mechanisms. The situation is further complicated by indications that tryptophan may in some circumstances be contraindicated in pain. There is evidence from work on animals that an ascending serotonergic system in the brain can counteract morphine analgesia. Thus, tryptophan antagonized morphine analgesia in the rat as measured by the formalin test (Abbott and Young 1989), which may be a model of tissue injury pain in humans. In a recent study tryptophan or placebo was given intravenously to patients receiving morphine after abdominal surgery

(Franklin et al. 1990). There was a trend for the tryptophan-treated group to have higher pain scores, and in the control group there was a strong positive relationship between plasma tryptophan and morphine requirements. This finding suggests that the decline in tryptophan levels that occurs after surgery may have potentiated morphine analgesia.

So far, no two studies on the analgesic effect of tryptophan have used the same type of patients. As confirmation of a clinically useful effect is always needed before it can be considered proven, confusion concerning the types of pain that may usefully be treated with tryptophan is unlikely to be resolved soon.

CONCLUSIONS AND PROSPECTS FOR THE FUTURE

Thirty years of work on the clinical psychopharmacology of tryptophan has resulted in surprisingly few firm conclusions, partly because of a proliferation of small studies, sometimes without adequate control groups, and a paucity of placebo-controlled studies of adequate size. However, even large placebo-controlled studies may not produce conclusive results if insufficient attention is paid to the way tryptophan is used and to the type of patient treated.

In sleep studies, the suggestion that the onset of tryptophan's action occurs only over several days needs to be tested adequately. If it is true, then a comparison of continuous use and interval therapy will be needed. So far, the majority of studies on the hypnotic effect of tryptophan have looked at only 1 or 2 nights. With longer studies a useful hypnotic effect in severe insomnia may be established.

In studies on depressed patients, the way tryptophan is given is, with a few exceptions, adequate. A single daily dose will ensure that serotonin synthesis is not elevated for a large portion of any day, and a divided dosage schedule is usually used. The primary problem in studies on affective disorders has been inadequate group size and lack of placebo control. The type of patient may also be important, as current evidence indicates that mildly or moderately depressed patients respond better than severely depressed patients. If the suggestion made in this chapter concerning arousal is correct, then one could predict that agitated depressed patients will respond better to tryptophan than retarded depressed patients, and that factors that arouse depressed patients, such as an exercise program, may potentiate the action of tryptophan.

The large number of studies on the action of tryptophan on depression and sleep were the result of early clinical observations that tryptophan influenced mood and arousal (Lauer et al. 1958; Smith and Prockop 1962). Less attention has been paid to the effect of tryptophan on pain and aggression. The stimulus for these studies came from animal work on the effects of serotonin on pain and aggression. Further animal work may help clinical progress in these areas. For example, a study on the use of tryptophan in different pain models may provide clues to the type(s) of clinical pain that will respond best to tryptophan.

In spite of the confusion surrounding some aspects of tryptophan use, there is good evidence that it does have therapeutic effects in some circumstances. The combination of these therapeutic effects and the relative lack of side effects makes tryptophan a useful drug. Further research should improve knowledge of the conditions in which tryptophan has beneficial therapeutic effects.

REFERENCES

Abbott FV, Young SN: Effect of 5-hydroxytryptamine precursors on morphine analgesia in the formalin test. Pharmacol Biochem Behav 31:855–860, 1989

Altman PL, Dittmer DS (eds): Metabolism. Bethesda, MD, Federal American Societies of Experimental Biology, 1968, pp 114–115

Anderson GM, Teff KL, Young SN: Serotonin in cisternal cerebrospinal fluid of the rat: measurement and use as an index of functionally active serotonin. Life Sci 40:2253–2260, 1987

Åsberg M, Schalling D, Traskman-Bendz L, et al: Psychobiology of suicide, impulsivity, and related phenomena, in Psychopharmacology: The Third Generation of Progress. Edited by Meltzer HY. New York, Raven Press, 1987, pp 655–668

Ayuso Gutierrez JL, Lopez-Ibor Alino JJ: Tryptophan and an MAOI (nialamide) in the treatment of depression: a double-blind study. International Pharmacopsychiatry 6:92–97, 1971

Baldessarini RJ: Treatment of depression by altering monoamine metabolism: precursors and metabolic inhibitor. Psychopharmacol Bull 20:224–239, 1984

Beitman BD, Dunner DL: L-Tryptophan in the maintenance treatment of bipolar II manic-depressive illness. Am J Psychiatry 139:1498–1499, 1982

Bender DA, Totoe L: High doses of vitamin B_6 in the rat are associated with inhibition of hepatic tryptophan metabolism and increased uptake of tryptophan into the brain. J Neurochem 43:733–736, 1984

Bennie EH: Mianserin hydrochloride and L-tryptophan compared in depressive illness. Br J Clin Social Psychiatry 1:90–91, 1982

Bowers MB: Cerebrospinal fluid 5-hydroxyindoles and behavior after L-tryptophan and pyridoxine administration to psychiatric patients. Neuropharmacology 9:599–604, 1970

Brewerton TD, Reus VI: Lithium carbonate and L-tryptophan in the treatment of bipolar and schizoaffective disorders. Am J Psychiatry 140:757–760, 1983

Broadhurst AD: L-Tryptophan versus E.C.T. Lancet 1:1392–1393, 1970

Bryan DJ: The role of urinary tryptophan metabolites in the etiology of bladder cancer. Am J Clin Nutr 24:841–846, 1971

Carroll BJ, Mowbray RM, Davies B: Sequential comparison of L-tryptophan with E.C.T. in severe depression. Lancet 1:967–969, 1970

Chambers CA, Naylor GJ: A controlled trial of L-tryptophan in mania. Br J Psychiatry 132:555–559, 1978

Chouinard G, Jones BD, Young SN, et al: Potentiation of lithium by tryptophan in a case of bipolar illness. Am J Psychiatry 136:719–720, 1979a

Chouinard G, Young SN, Annable L, et al: Tryptophan-nicotinamide, imipramine and their combination in depression: a controlled study. Acta Psychiatr Scand 59:395–414, 1979b

Chouinard G, Young SN, Bradwejn J, et al: Tryptophan in the treatment of depression and mania, in Management of Depressions with Monoamine Precursors: Advances in Biological Psychiatry, Vol 10. Edited by van Praag HM, Mendlewicz J. Basel, Karger, 1983, pp 47–66

Chouinard G, Young SN, Annable L: A controlled clinical trial of L-tryptophan in acute mania. Biol Psychiatry 20:546–557, 1985

Cole JO, Hartmann E, Brigham P: L-Tryptophan: clinical studies, in Psychopharmacology Update. Edited by Cole JO. Lexington, MA, The Collamore Press, 1980, pp 119–148

Cooper AJ, Datta SR: A placebo controlled evaluation of L-tryptophan in depression in the elderly. Can J Psychiatry 25:336–390, 1980

Coppen A, Shaw DM, Farrell JP: Potentiation of the antidepressant effect of a monoamine-oxidase inhibitor by tryptophan. Lancet: 1:79–81, 1963

Coppen A, Shaw DM, Herzberg B, et al: Tryptophan in the treatment of depression. Lancet 2:1178–1180, 1967

Coppen A, Whybrow PC, Noguera R, et al: The comparative antidepressant value of L-tryptophan and imipramine with and without attempted potentiation by liothyronine. Arch Gen Psychiatry 26:234–241, 1972

d'Elia G, Lehmann J, Raotma H: Evaluation of the combination of tryptophan and ECT in the treatment of depression. I. Clinical analysis. Acta Psychiatr Scand 56:303–318, 1977

Eccleston D, Ashcroft GW, Crawford TBB, et al: Effect of tryptophan administration on 5-HIAA in cerebrospinal fluid in man. J Neurol Neurosurg Psychiatry 33:269–272, 1970

Farkas T, Dunner DL, Fieve RR: L-Tryptophan in depression. Biol Psychiatry 11:295–302, 1976

Fernstrom JD, Faller DV: Neutral amino acids in the brain: changes in response to food ingestion. J Neurochem 30:1513–1538, 1978

Fernstrom JD, Wurtman RJ: Brain serotonin content: increase following ingestion of a carbohydrate diet. Science 174:1023–1025, 1971

Franklin KBJ, Abbott FV, English MJM, et al: Tryptophan-morphine interactions and postoperative pain. Pharmacol Biochem Behav 35:157–163, 1990

Gallagher DW, Aghajanian GK: Inhibition of firing of raphe neurons by tryptophan and 5-hydroxytryptophan: blockade by inhibiting serotonin synthesis with Ro-4-1602. Neuropharmacology 15:149–156, 1976

Gayford JJ, Parker AL, Phillips EM, et al: Whole blood 5-hydroxytryptamine during treatment of endogenous depressive illness. Br J Psychiatry 122:597–598, 1973

Gelenberg AS, Gibson CJ, Wojcik JD: Neurotransmitter precursors for the treatment of depression. Psychopharmacol Bull 18:7–18, 1982

Gillin JC, Kaplan JA, Wyatt RJ: Clinical effects of tryptophan in chronic schizophrenic patients. Biol Psychiatry 11:635–639, 1976

Glassman AH, Platman SR: Potentiation of a monoamine oxidase inhibitor by tryptophan. J Psychiatr Res 7:83–88, 1969

Gnirss F, Schneider-Helmert D, Schenker J: L-Tryptophan + oxprenolol: a new approach to the treatment of insomnia. Pharmacopsychiatria 11:180–185, 1978

Grahame-Smith DG: Studies in vivo on the relationship between tryptophan, brain 5-HT synthesis and hyperactivity in rats treated with a monoamine oxidase inhibitor and L-tryptophan. J Neurochem 18:1053–1066, 1971

Green AR, Aronson JK: Metabolism of an oral tryptophan load. III. Effect of a pyridoxine supplement. Br J Clin Pharmacol 10:617–619, 1980

Green AR, Aronson JK, Curzon G, et al: Metabolism of an oral tryptophan load. II. Effects of pretreatment with the putative tryptophan pyrrolase inhibitors nicotinamide or allopurinol. Br J Pharmacol 10:611–615, 1980

Greenwald BS, Marin DB, Silverman SM: Serotonnergic treatment of screaming and banging in dementia. Lancet 2:1464–1465, 1986

Hale AS, Procter AW, Bridges PK: Clomipramine, tryptophan and lithium in combination for resistant endogenous depression: seven case studies. Br J Psychiatry 151:213–217, 1987

Hankes LV, Brown RR, Leklem J, et al: Metabolism of ^{14}C labeled enantiomers of tryptophan, kynurenine and hydroxykynurenine in humans with scleroderma. J Invest Dermatol 58:85–95, 1972

Harris B: Prospective trial of L-tryptophan in maternity blues. Br J Psychiatry 137:233–235, 1980

Hartmann E: Possible effects of tryptophan ingestion. J Nutr 117:1314, 1987

Hartmann E, Greenwald D: Tryptophan and human sleep: an analysis of 43 studies, in Progress in Tryptophan and Serotonin Research. Edited by Schlossberger HG, Kochen W, Linzen B, et al. Berlin, Walter de Gruyter, 1984, pp 297–304

Hartmann E, Spinweber CL: Sleep induced by L-tryptophan: effect of dosages within the normal dietary intake. J Nerv Ment Dis 167:497–499, 1979

Hartmann E, Cravens J, List S: Hypnotic effects of L-tryptophan. Arch Gen Psychiatry 31:394–397, 1974

Hartmann E, Spinweber CL, Ware C: L-Tryptophan, L-leucine, and placebo: effects on subjective alertness. Sleep Res 5:57, 1976

Hartmann E, Lindsley JG, Spinweber C: Chronic insomnia: effects of tryptophan, flurazepam, secobarbital, and placebo. Psychopharmacology (Berlin) 80:138–142, 1983

Hattori M, Kotake Y, Kotake Y, et al: Studies on the urinary excretion of xanthurenic acid in diabetics, in Progress in Tryptophan and Serotonin Research. Edited by Schlossberger HG, Kochan W, Linzen B, et al. Berlin, Walter de Gruyter, 1984, pp 347–354

Hayaishi O: Indoleamine-2,3-dioxygenase, with special reference to the mechanism of interferon action. Biken J 28:39–49, 1985

Herrington RN, Bruce A, Johnstone EC, et al: Comparative trial of L-tryptophan and amitriptyline in depressive illness. Psychol Med 6:673–678, 1974

Herrington RN, Bruce A, Johnstone EC, et al: Comparative trial of L-tryptophan and E.C.T. in severe depressive illness. Lancet 2:731–734, 1976

Hertz D, Sulman FG: Preventing depression with tryptophan. Lancet 1:531–532, 1968

Hess SM, Doepfner W: Behavioral effects and brain amine content in rats. Arch Int Pharmacodyn Ther 134:89–99, 1961

Honore P, Møller SE, Jorgensen A: Lithium + L-tryptophan compared with amitriptyline in endogenous depression. J Affective Disord 4:79–82, 1982

Hosobuchi Y: Tryptophan reversal of tolerance to analgesia induced by central grey stimulation. Lancet 2:47, 1978

Hosobuchi Y, Lamb S, Baskin D: Tryptophan loading may reverse tolerance to opiate analgesics in humans: a preliminary report. Pain 9:161–169, 1980a

Hosobuchi Y, Rossier J, Bloom FE: Oral loading with L-tryptophan may augment the simultaneous release of ACTH and beta-endorphin that accompanies periaqueductal stimulation in humans, in Neural Peptides and Neuronal Communications. Edited by Costa E, Trabucchi M. New York, Raven Press, 1980b, pp 563–570

Ikeda S, Kotake Y: Urinary excretion of xanthurenic acid and zinc in diabetes, in Progress in Tryptophan and Serotonin Research. Edited by Schlossberger HG, Kochen W, Linzen B, et al. Berlin, Walter de Gruyter, 1984, pp 355–358

King RB: Pain and tryptophan. J Neurosurg 53:44–52, 1980

Kirkegaard C, Møller SE, Bjorum N: Addition of L-tryptophan to electroconvulsive treatment in endogenous depression: a double-blind study. Acta Psychiatr Scand 58:457–462, 1978

Kline NS, Shah BK: Comparable therapeutic efficacy of tryptophan and imipramine: average therapeutic ratings versus "true" equivalence: an important difference. Curr Ther Res 15:484–487, 1973

Lauer JW, Inskip WM, Bernsohn J, et al: Observations on schizophrenic patients after iproniazid and tryptophan. AMA Arch Neurol Psychiatry 80:122–130, 1958

Lieberman HR, Corkin S, Spring BJ, et al: Mood, performance, and pain sensitivity: changes induced by food constituents. J Psychiatr Res 17:135–145, 1983

Lindberg D, Ahlfors UG, Dencker SJ, et al: Symptom reduction in depression after treatment with L-tryptophan or imipramine: item analysis of Hamilton rating scale for depression. Acta Psychiatr Scand 60:287–294, 1979

Linnoila M, Viukari M, Nummunen A, et al: Efficacy and side effects of chloral hydrate and tryptophan as sleeping aids in psychogeriatric patients. International Pharmacopsychiatry 15:124–128, 1980

Lopez-Ibor Alino JJ, Ayuso Gutierrez JL, Montejo Iglesias ML: Tryptophan and amitriptyline in the treatment of depression: a double-blind study. International Pharmacopsychiatry 8:145–151, 1973

MacSweeney DA: Treatment of unipolar depression. Lancet 2:510–511, 1975

Meier AH, Wilson JM: Tryptophan feeding adversely influences pregnancy. Life Sci 32:1193–1196, 1983

Mendels J, Stinnett JL, Burnes D, et al: Amine precursors and depression. Arch Gen Psychiatry 32:22–30, 1975

Mitchell MJ, Daines GE, Thomas BL: Effect of L-tryptophan and phenylalanine on burning pain threshold. Phys Ther 67:203–205, 1987

Moldofsky H, Lue FA: The relationship of alpha and delta EEG frequencies to pain and mood in "fibrositis" patients treated with chlorpromazine and L-tryptophan. Electroencephalogr Clin Neurophysiol 50:1–80, 1980

Møller SE: Pharmacokinetics of tryptophan, renal handling of kynurenine and the effect of nicotinamide on its appearance in plasma and urine following L-tryptophan loading of healthy subjects. Eur J Clin Pharmacol 21:137–142, 1981

Møller SE, Kirk L: The effect of allopurinol on the kynurenine formation in humans following a tryptophan load. Acta Vitaminol Enzymol 32:159–162, 1978

Morand C, Young SN, Ervin FR: Clinical response of aggressive schizophrenics to oral tryptophan. Biol Psychiatry 18:575–578, 1983

Murphy DL, Baker M, Goodwin FK, et al: L-Tryptophan in affective disorders: indoleamine changes and differential clinical effects. Psychopharmacology (Berlin) 34:11–20, 1974

National Cancer Institute: Bioassay of L-tryptophan for possible carcinogenicity. National Cancer Institute Carcinogenesis Technical Report Series No. 71 (DHEW Publication No. [NIH] 78-1321). Washington, DC, US Government Printing Office, 1978

Nurmikko T, Pertovaara A, Pontinen PJ, et al: Effect of L-tryptophan supplementation on ischemic pain. Acupunct Electrother Res 9:44–55, 1984

Oates JA, Sjoerdsma A: Neurologic effects of tryptophan in patients receiving a monoamine oxidase inhibitor. Neurology 10:1076–1078, 1960

Oldendorf WH, Szabo J: Amino acid assignment to one of three blood-brain barrier amino acid carriers. Am J Physiol 230:94–98, 1976

Olson RE, Gursey D, Vester JW: Evidence for a defect in tryptophan metabolism in chronic alcoholism. N Engl J Med 263:1169–1174, 1960

O'Neil M, Page N, Adkins WN, et al: Tryptophan/trazodone treatment of aggressive behavior. Lancet 2:859–860, 1986

Pare CMB: Potentiation of monoamine-oxidase inhibitors by tryptophan. Lancet 2:527–528, 1963

Perez-Cruet J, Chase TN, Murphy DL: Dietary regulation of brain tryptophan metabolism by plasma ratio of free tryptophan and neutral amino acids in humans. Nature 248:693–695, 1974

Prange AJ, Wilson IC, Lynn CW, et al: L-Tryptophan in mania: contribution to a permissive hypothesis of affective disorders. Arch Gen Psychiatry 30:56–62, 1974

Rao B, Broadhurst AD: Tryptophan and depression. Br Med J 2:460, 1976

Sangdee C, Franz DN: Lithium enhancement of central 5HT transmission induced by 5HT precursors. Biol Psychiatry 15:59–75, 1980

Schneider-Helmert D: Interval therapy with L-tryptophan in severe chronic insomniacs: a predictive laboratory study. International Pharmacopsychiatry 16:162–173, 1981

Schneider-Helmert D, Spinweber CL: Evaluation of L-tryptophan for treatment of insomnia: a review. Psychopharmacology (Berlin) 89:1–7, 1986

Seltzer S, Stoch R, Marcus R, et al: Alteration of human pain thresholds by nutritional manipulation and L-tryptophan supplementation. Pain 13:385–393, 1982

Seltzer S, Dewart D, Pollack R, et al: The effects of dietary tryptophan on chronic maxillofacial pain and experimental pain tolerance. J Psychiatr Res 17:181–186, 1983

Shaw DM, MacSweeney DA, Hewland R, et al: Tricyclic antidepressants and tryptophan in unipolar depression. Psychol Med 5:276–278, 1975

Shopsin B: Enhancement of the antidepressant response to L-tryptophan by a liver pyrrolase inhibitor: a rational treatment approach. Neuropsychobiology 4:188–192, 1978

Shpeen SE, Morse DR, Furst ML: The effect of tryptophan on postoperative endodontic pain. Oral Surgery 58:446–449, 1984

Smith B, Prockop DJ: Central-nervous-system effects of ingestion of L-tryptophan by normal subjects. N Engl J Med 267:1338–1341, 1962

Sourkes TL: Toxicology of monoamine precursors, in Advances in Biological Psychiatry, Vol 10. Edited by van Praag HM, Mendlewicz J. Basel, Karger, 1983, pp 160–175

Spinweber CL: L-Tryptophan administered to chronic sleep-onset insomniacs: late-appearing reduction of sleep latency. Psychopharmacology (Berlin) 90:151–155, 1986

Sternbach RA, Janowsky DS, Huey LY, et al: Effects of altering brain serotonin activity on human chronic pain, in Advances in Pain Research and Therapy. Edited by Bonica JJ, Albe-Fessard D. New York, Raven Press, 1976, pp 601–606

Teff KL, Young SN, Marchand L, et al: Acute effect of protein or carbohydrate breakfasts on human cerebrospinal fluid monoamine precursor and metabolite levels. J Neurochem 52:235–241, 1989

Thomson J, Rankin H, Ashcroft GW, et al: The treatment of depression in general practice: a comparison of L-tryptophan, amitriptyline, and a combination of L-tryptophan and amitriptyline with placebo. Psychol Med 12:741–751, 1982

Trulson ME, Jacobs BL: Dose-response relationships between systematically administered L-tryptophan or L-5-hydroxytryptophan and raphe unit activity in the rat. Neuropharmacology 15:339–344, 1976

Trulson ME, Jacobs BL: Raphe unit activity in freely moving cats: correlations with level of behavioral arousal. Brain Res 169:135–150, 1979

Trulson ME, Sampson HW: Ultrastructural changes of the liver following L-tryptophan ingestion in rats. J Nutr 116:1109–1115, 1986

van Praag HM, Flentge F, Korf J, et al: The influence of probenecid on the metabolism of serotonin, dopamine and their precursors in man. Psychopharmacology (Berlin) 33:141–151, 1973

Walinder J, Skott A, Carlsson A, et al: Potentiation of the antidepressant action of clomipramine by tryptophan. Arch Gen Psychiatry 33:1384–1389, 1976

Walinder J, Carlsson A, Persson R: 5-HT reuptake inhibitors plus tryptophan in endogenous depression. Acta Psychiatr Scand 63 (suppl 290):179–190, 1981

Watson AJS, Phillips K: Serotoninergic treatment of screaming and banging in dementia. Lancet 2:1464–1465, 1986

Wilcock GK, Stevens J, Perkins A: Trazodone/tryptophan for aggressive behavior. Lancet 1:929–930, 1987

Worrall EP, Moody JP, Peet M, et al: Controlled studies of the acute antidepressant effects of lithium. Br J Psychiatry 135:255–262, 1979

Wurtman RJ, Hefti F, Melamed E: Precursor control of neurotransmitter synthesis. Pharmacol Rev 32:315–335, 1981

Wyatt RJ, Engelman K, Kupfer DJ, et al: Effects of L-tryptophan (a natural sedative) on human sleep. Lancet 2:842–846, 1970

Yokogoshi H, Wurtman RJ: Meal composition and plasma amino acid ratios: effect of various proteins or carbohydrates, and of various protein concentrations. Metabolism 35:837–842, 1986

Young SN: The clinical psychopharmacology of tryptophan, in Nutrition and the Brain, Vol 7. Edited by Wurtman RJ, Wurtman JJ. New York, Raven Press, 1986, pp 49–88

Young SN, Gauthier S: Effect of tryptophan administration on tryptophan, 5-hydroxyindoleacetic acid and indoleacetic acid in human lumbar and cisternal cerebrospinal fluid. J Neurol Neurosurg Psychiatry 44:323–328, 1981

Young SN, Oravec M: The effect of growth hormone on the metabolism of a tryptophan load in liver and brain of hypophysectomized rats. Canadian Journal of Biochemistry 57:517–522, 1979

Yuwiler A, Brammer GL, Morley JE, et al: Short-term and repetitive administration of oral tryptophan in normal men. Arch Gen Psychiatry 38:619–626, 1981

Chapter 4

Catecholamine Precursor Research in Depression: The Practical and Scientific Yield

Herman M. van Praag, M.D., Ph.D.

Chapter 4

Catecholamine Precursor Research in Depression: The Practical and Scientific Yield

The classical antidepressants increase the availability of serotonin (5-hydroxytryptamine [5-HT]) and noradrenaline (NA) in the central nervous system (CNS): tricyclic antidepressants by inhibiting their uptake and monoamine oxidase inhibitors (MAOIs) by inhibiting their degradation. This increase is considered instrumental to the therapeutic effect of these drugs. These observations prompted the search for disturbances in metabolism and function of monoamines in depression, in particular in those syndromes responsive to antidepressant treatment.

Precursors of monoamines have been used to enhance production and release of these amines in the CNS. This is a valid strategy. Loading with both tyrosine and dopa leads to a rise in the concentration of homovanillic acid (HVA) and 3-methoxy-4-hydroxyphenylglycol (MHPG) in cerebrospinal fluid (CSF); loading with the 5-HT precursors tryptophan and 5-hydroxytryptophan (5-HTP) leads to an increasing concentration of CSF 5-hydroxyindoleacetic acid (5-HIAA) (van Praag and Lemus 1986). HVA, MHPG, and 5-HIAA are the major metabolites of dopamine, NA, and 5-HT, and a rise in concentration of these substances reflects an augmented synthesis and release of dopamine, NA, and 5-HT respectively in the CNS. So far, two monoamine precursors have been found to have general therapeutic applications—dopa in Parkinson's disease and 5-HTP in myoclonus and (at least in certain European countries) in depression.

STUDIES WITH DOPA AND TYROSINE

Van Praag and Korf (1971) reported the occurrence of dopamine disturbances in depression, most notably in vital depression (major depression, melancholic type, in this chapter called melancholic

79

depression). The authors used the probenecid technique. Probenecid blocks the efflux of HVA (and 5-HIAA) from the CNS (including the CSF) to the bloodstream. The magnitude of the accumulation of these metabolites in the CSF after probenecid loading is a crude indication of the metabolism of the mother amines in the CNS. In some patients with vital depression, the probenecid-induced accumulation of HVA, the major metabolite of dopamine, in CSF was decreased, indicating a lowered dopamine metabolism in the CNS, particularly in the nigrostriatal dopamine system, the region that is predominantly represented by CSF HVA. Guided by Parkinson's disease, with its severe disturbances in motor functioning and profound dopamine depletion, we searched for correlations between lowered CSF HVA and motor retardation. We noted that in comparing groups of depressed patients with and without pronounced motor retardation and lack of initiative, the HVA accumulation after probenecid appeared to be half as high in the retarded patients as in the nonretarded depressives and in a control group (Figure 4-1).

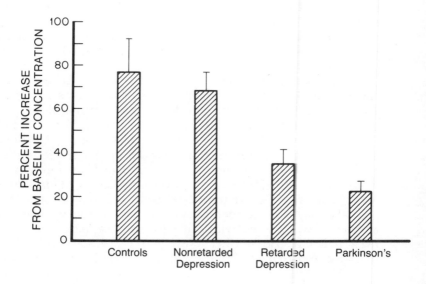

Figure 4-1.   Post-probenecid concentration of HVA in CSF in controls ($n = 12$), depression with ($n = 11$) and without ($n = 13$) motor retardation, and in Parkinson's disease ($n = 14$). Reprinted from van Praag and Korf 1971 and Lakke et al. 1972, with permission of Springer-Verlag and Macmillan, respectively. Copyright 1971 and 1972.

Similar findings were reported by other investigators (Banki 1977; Banki et al. 1981; Papeschi and McClure 1971).

If lowered dopamine metabolism in the nigrostriatal system underlies retardation and inertia, increasing dopamine availability could be expected to exert an activating and energizing effect. L-Dopa was our drug of choice to booster dopamine metabolism, since we had demonstrated in humans that it increases dopamine metabolism substantially, while having little impact on NA metabolism. Moreover, the increase in dopamine metabolism persists in time (van Praag and Lemus 1986) (Figure 4-2).

L-Dopa (average 260 mg/day) in combination with a peripheral decarboxylase inhibitor was first studied in a group of unselected melancholically depressed patients and found to be devoid of overall therapeutic effects. Subsequently, we compared the effect of L-dopa (290 mg/day) and a peripheral decarboxylase inhibitor in melancholic patients with lowered post-probenecid CSF HVA and in melancholic patients with normal HVA response. Motor retardation

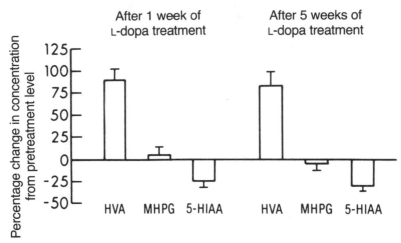

Figure 4-2. L-Dopa (mean, 290 mg/day) in combination with the peripheral decarboxylase inhibitor MK 486 (150 mg/day) induced in five test persons a significant increase in post-probenecid CSF HVA. The concentration of MHPG did not change significantly, while post-probenecid 5-HIAA decreased slightly but consistently (in all test persons). CSF concentrations were measured before treatment and after 1 week and 5 weeks of L-dopa treatment. The changes persisted over time. Reprinted from van Praag and Lemus 1986, with permission of Raven Press. Copyright 1986.

and inertia were significantly more pronounced in the former group. L-Dopa was shown to improve motor functioning and level of initiative in the low-HVA group and to normalize the HVA level (Figure 4-3). Mood and hedonic functioning were not significantly influenced. Anxiety levels rose slightly, but significantly. In the group with normal HVA response, L-dopa failed to exercise therapeutic effects (van Praag and Korf 1975).

The latter experiment was repeated in a different group of patients using L-tyrosine instead of L-dopa (van Praag 1986). Unlike L-dopa, L-tyrosine, the precursor of the catecholamines dopamine and NA, significantly increases CSF MHPG in humans, indicating risen NA metabolism and release. CSF HVA, on the other hand, rises only slightly, indicating that tyrosine's impact on dopamine metabolism (in the nigrostriatal system) is less pronounced than that of L-dopa (Figure 4-4).

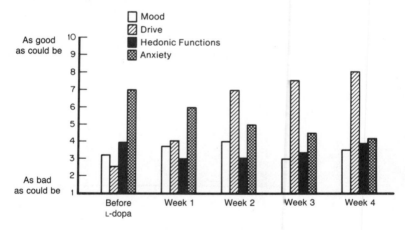

Figure 4-3.   L-Dopa (average 260 mg/day) in combination with a peripheral decarboxylase inhibitor was administered for 4 weeks to 10 patients with major depression, melancholic type, pronounced motor retardation, and low post-probenecid CSF HVA. Motor retardation and inertia ('drive') improved significantly. Mood and hedonic functioning were not significantly influenced. Post-probenecid CSF HVA was normalized. In melancholic patients with normal HVA response and without motor retardation, L-dopa was ineffectual and no better than placebo. Reprinted from van Praag and Korf 1975 and van Praag and Lemus 1986, with permission of Georg Thieme Verlag and Raven Press, respectively. Copyright 1975 and 1986.

Again studying melancholic patients with and without lowered CSF HVA accumulation after probenecid, we observed no therapeutic effect of (100 mg/kg/day) L-tyrosine on motor retardation and inertia in either group (Figures 4-5 and 4-6). Normalization of CSF HVA did not occur either.

Based on these findings, we hypothesized that diminished dopamine metabolism in the nigrostriatal system underlies decreased motor activity and lowered level of initiative, irrespective of nosological diagnosis (van Praag et al. 1975). In accord with this hypothesis is the observation by Lindstrom (1985) that low CSF HVA occurs in schizophrenic patients with pronounced lassitude and slowness of movement.

Linking the dopamine system to the ability to carry out goal-directed behavior, one must take into account that such behavior is a composite of different components. First, an initial stimulus is needed.

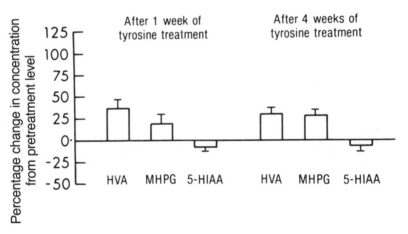

Figure 4-4.   L-Tyrosine (100 mg/kg/day) induced in five test persons a significant increase in CSF MHPG. CSF HVA increased significantly as well, but the increase was much less pronounced than after L-dopa administration (see Figure 4-2). CSF 5-HIAA did not change significantly. CSF concentrations were measured before treatment and after 1 and 4 weeks of treatment. The changes persisted over time. CSF concentrations were measured after probenecid loading. Lumbar punctures were performed 2 hours after a dose of tyrosine. Reprinted from van Praag and Lemus 1986, with permission of Raven Press. Copyright 1986.

This stimulus can be an instinctual drive, such as hunger, thirst, or sex; an emotion, such as anger; or a cognitive set, such as the realization of having to prepare for a test. Next, the goals for responding appropriately to the stimulus have to be selected—e.g., find food to satisfy hunger; display aggressive behavior to reduce anger; move to a library and study for a test. Subsequently, the behavior to attain these goals has to be initiated and sustained. Finally, signals need to be set in operation so that when the mission has been accomplished, the behavior can be terminated. The term "drive" is used to indicate an initial stimulus of an instinctual nature. It is also used for the entire process of initiation and completion of goal-directed behavior. For the sake of brevity, I use the term in this chapter in the latter sense.

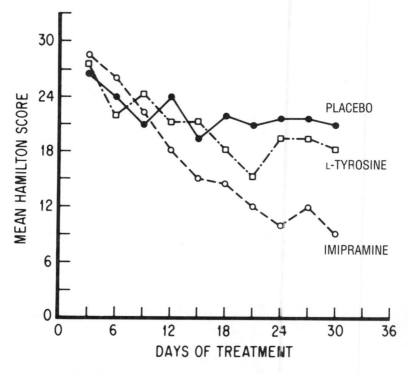

Figure 4-5. L-Tyrosine (100 mg/kg/day), imipramine (225 mg/day), and placebo were double blindly compared in 3 groups of 10 patients each with major depression, melancholic type. The overall therapeutic yield of tyrosine was no better than that of placebo.

The sequence leading to goal-directed behavior can be disturbed at any juncture. The human data suggest that dopamine is involved in the mobilization, facilitation, and sustenance of goal-directed behavior. The large pool of animal data available concurs with this conclusion (Ashton 1987; Crow and Deakin 1985; Freed and Yamamoto 1985).

STUDIES WITH TYROSINE AND 5-HT PRECURSORS

Several controlled and uncontrolled studies indicate that 5-HTP has antidepressant potential while tryptophan does not or does to a much lesser degree (van Praag 1981). In a placebo-controlled, comparative study of L-tryptophan and L-5-HTP, I confirmed the observations obtained in studies with each precursor separately (Figure 4-7) (van Praag 1984). I searched for an explanation for the discrepant findings with the two 5-HT precursors. It is apparent that both increase central 5-HT metabolism to an equal extent, as reflected in the probenecid-induced accumulation of 5-HIAA in CSF. They appeared to differ,

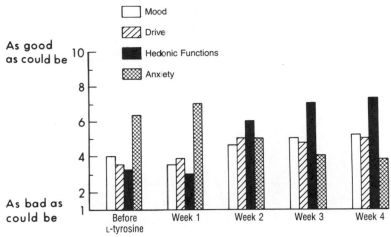

Figure 4-6. L-Tyrosine (100 mg/kg/day) was administered for 4 weeks to 10 patients with major depression, melancholic type, pronounced motor retardation, and low post-probenecid CSF HVA. Tyrosine had no effect on motor retardation and inertia ("drive"); neither did placebo. Mood and anxiety did not change significantly. Hedonic functions improved significantly (see also Figure 4-13). Normalization of post-probenecid CSF HVA did not occur.

however, in their effect on catecholamine metabolites in CSF (reflecting intracerebral metabolism of the mother amines) 5-HTP increases the metabolism of dopamine as well as of NA; tryptophan does not (van Praag 1983) (Figure 4-8). In high doses (> 5 g iv), tryptophan even lowers dopamine and NA metabolism (Figure 4-9), possibly by interfering with the influx of tyrosine, the mother substance of catecholamine, in the CNS (van Praag et al. 1987a).

The ability of 5-HTP to stimulate catecholamine metabolism is thought to be due to the presence of aromatic amino acid decarboxylase in catecholaminergic nerve cells. Hence, 5-HTP is converted in 5-HT not only in 5-HT-ergic but also in catecholaminergic neurons. In the latter, 5-HT functions as a false transmitter, causing

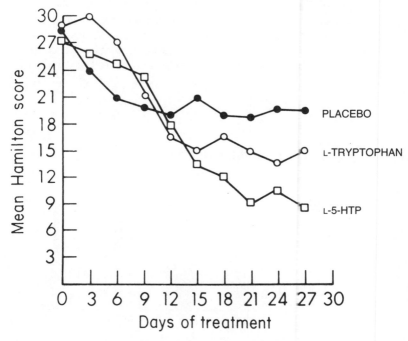

Figure 4-7. Comparative, controlled study of 5-HTP (200 mg/day) in combination with carbidopa (150 mg/day) (□), L-tryptophan (5 g/day) (○), and placebo (●) in patients suffering from major depression, melancholic type. 5-HTP is significantly superior to tryptophan and placebo. Tryptophan treatment does not differ significantly from placebo treatment. Reprinted from van Praag 1984, with permission of *Psychopharmacology Bulletin*. Copyright 1984.

the synthesis rate of catecholamine to increase. The net functional effect of the two opposing processes, i.e., false transmitter formation leading to decreased function and increased catecholamine production leading to augmented function, is probably heightened catecholaminergic activity.

If 5-HTP's therapeutic superiority to tryptophan is related to its combined effects on 5-HT and catecholamine systems, one would expect tryptophan's antidepressant efficacy to be raised above the significance level by combining it with a compound capable of increasing catecholamine availability. Tyrosine is such a compound, as I showed in humans, and combining tryptophan with tyrosine indeed led to significant antidepressant activity (van Praag 1986) (Figure 4-10).

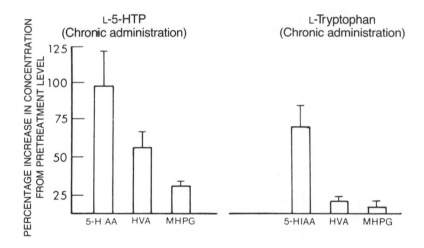

Figure 4-8. Percentage increase in the concentration of 5-HIAA, HVA, and MHPG in lumbar CSF after oral administration to test subjects of 200 mg L-5-HTP per day for 1 week (in combination with a peripheral decarboxylase inhibitor) or of L-tryptophan, 5 g per day for 1 week. After both 5-HT precursors, CSF 5-HIAA rose significantly. Catecholamine metabolite concentrations did not change after tryptophan and rose significantly after 5-HTP. Lumbar CSF was withdrawn after probenecid loading (5 g/5 hours). Reprinted from van Praag 1983, with permission of Pergamon Press. Copyright 1983.

A second observation supports the hypothesis that 5-HTP derives its therapeutic potential from its dual influence on 5-HT-ergic and catecholaminergic systems. In about 25% of patients initially treated successfully with 5-HTP, the response subsided in the second month of treatment (van Praag 1983). This phenomenon is paralleled by normalization of catecholamine metabolism (again, as reflected in the CSF concentration of catecholamine metabolites) while the metabolism of 5-HT remains increased (Figure 4-11). I hypothesized that clinical relapse and normalization of catecholamine metabolism were related. If so, reactivating catecholamine metabolism should restore clinical remission and, indeed, it did. Adding tyrosine to the 5-HTP regime in the group of relapsers once again led to subsidence of the depression and to a rise of catecholamine metabolism (Figure 4-12).

These data suggest that *combined* augmentation of 5-HT and catecholamine availability in the CNS provides the best conditions for antidepressant activity, and they argue against the existence of a separate diagnostic entity called "5-HT depression." The behavioral

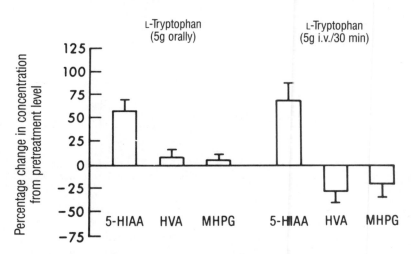

Figure 4-9.   Percentage increase in the concentration of 5-HIAA, HVA, and MHPG in lumbar CSF after administration to five test subjects of a single dose of 5 g L-tryptophan either orally or intravenously. 5-HIAA concentration rose significantly on both occasions and to the same extent. After oral administration, changes in HVA and MHPG were insignificant; after intravenous administration, both concentrations dropped significantly. Lumbar CSF was withdrawn after probenecid loading (5 g/5 hours) 8 hours after starting the load.

effects of 5-HTP in monkeys have also been demonstrated to be a function of its influence on 5-HT and catecholamine systems (Raleigh 1987).

THERAPEUTIC EFFECTS OF TYROSINE IN DEPRESSION

In hospitalized depressed patients, as mentioned, I observed no overall therapeutic effects of L-tyrosine. Yet, the substance was not devoid of pharmacological actions. When its influence on components of the depressive syndrome was analyzed, we observed that whereas the effects on mood and drive were not significant, those on anxiety and hedonic functioning were. Anxiety ratings increased; those for anhedonia decreased (Figure 4-13). Though preliminary, the latter observation is remarkable in that it suggests involvement of an NA-ergic deficit in the hedonic dysfunction.

In animals, brain systems that code for naturally occurring rewarding and aversive experiences have been shown. The existence of the so-called reward and punishment systems was deduced from the

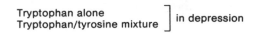

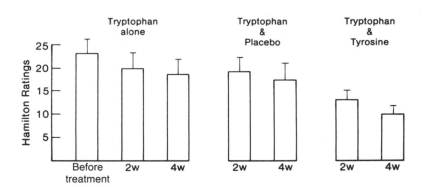

Figure 4-10. Average Hamilton Depression Rating Scale scores in 10 patients suffering from major depression, melancholic type. They were first treated for 4 weeks with L-tryptophan (5 g/day orally) and without therapeutic effects. Subsequently, in five of these patients L-tyrosine (100 mg/kg/day) was added to the tryptophan regime and in the five remaining patients placebo. The former combination led to significant improvement; the latter did not.

observation that rats with chronically implanted electrodes will work to obtain electric stimulation at specific sites in the brain, while they work to avoid stimulation at other specific sites (Olds and Milner 1954). Highest self-stimulation rates are obtained when electrodes are located in the medial forebrain bundle in the lateral hypothalamus.

Reward and punishment systems are, at least in part, monoaminergically innervated, the former by catecholamines, the latter by 5-HT. Dopamine is thought to be particularly involved in mobilizing and directing goal-directed behaviors. Dopamine neurons constitute a "go-system" or "incentive-system" guiding the organism to areas in

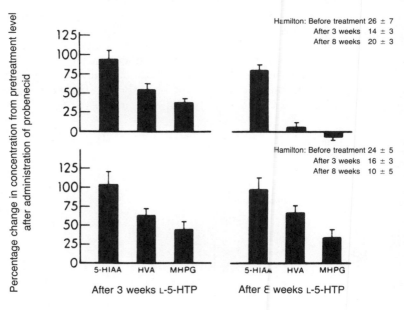

Figure 4-11. Augmenting effect of L-5-HTP (200 mg/day in combination with carbidopa, 150 mg/day) on catecholamine metabolism is transient in patients who relapsed in second month of 5-HTP treatment. In first month of treatment, there is a substantial increase of HVA and MHPG in CSF after a probenecid load. In second month, the levels of those metabolites have returned to normal. 5-HIAA response remains increased over time (top). In patients in whom the therapeutic response to 5-HTP continued, effect on catecholamine metabolites in CSF did not subside (bottom). Reprinted from van Praag 1983, with permission of Pergamon Press. Copyright 1983.

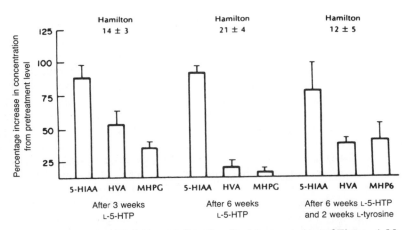

Figure 4-12. In the patients referred to in the top section of Figure 4-11, L-tyrosine (100 mg/kg/day) was added to the 5-HTP regime. This change led to significant clinical improvement and a significant rise of the CSF HVA and MHPG concentrations after probenecid. Reprinted from van Praag 1983, with permission of Pergamon Press. Copyright 1983.

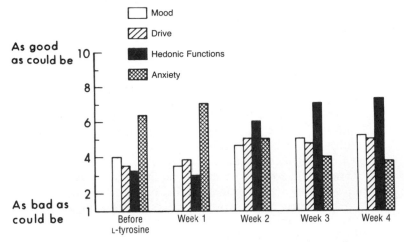

Figure 4-13. In 10 patients with major depression, melancholic type, L-tyrosine (100 mg/kg/day for 4 weeks) did not change the overall Hamilton Depression Rating Scale score. Discrete components of the syndrome, however, did change significantly. Hedonic functioning improved (see also Figure 4-6) while anxiety levels increased. Placebo did not change these psychopathological dimensions.

the environment associated with reward (Crow 1973). Though a role for NA in self-stimulation is less well supported than for dopamine, NA-ergic neurons seem to play a role in what I have called "emotional memory" (van Praag et al. 1988); that is, in the consolidation and retrieval of the emotional arousal induced by certain behaviors. Alternatively, NA-ergic tracts constitute the connections between the neural representation of rewards anticipated from particular behaviors and the dopaminergic incentive system (Crow 1973, 1977). NA-ergic mechanisms have also been implicated in selecting among a choice of possible behaviors to attain a particular goal, based on experiences of reward or punishment those behaviors elicited previously (Ashton 1987). A reward-related role of NA-ergic mechanisms in self-stimulation is supported by experiments with clonidine, an α_2-adrenergic agonist. This substance disrupts the rewarding component of brain stimulation in a selective manner, probably by activating presynaptic α_2-receptors and thus curtailing NA-ergic impulse flow (Franklin 1978; Hunt et al. 1976). The catecholamine synthesis inhibitor α-methyl-ρ-tyrosine magnifies this effect (Hunt et al. 1976).

NA-ergic mechanisms, then, seem to enable the animal to couple the experience of reward or anticipated reward to a particular activity. The hedonic component of the melancholic syndrome can be conceived of as an inability of exactly that coupling mechanism. "I used to love to go with my kids to a baseball game, Doctor, but somehow it does not get me excited anymore." Situations that used to create pleasure are perceived as usual but are now devoid of emotional charge. Stein (1978) was the first to suggest depression to be "a disorder of positive reinforcement or reward function." I do not subscribe to this generalization but would suggest a more focused version of this hypothesis. The hedonic disturbance in depression is, I submit, related to NA-ergic insufficiency, a transmission defect resulting in inability to couple the reward component to mental or physical activities that used to carry such a charge.

The NA hypothesis of anhedonia fits into the general framework of classic information processing theory of human cognition, in which two sets of structures are postulated: one for storing information, and the other for transferring information from one structure to another. In this model—and I follow here Kihlstrom's (1987) description—information from the environment is transduced into a pattern of neural impulses by the sensory receptors. This pattern is briefly held in the sensory registers, one for each modality, where it is then analyzed by processes known as feature detection and pattern recognition. Information that has been identified as meaningful and relevant to current goals is then transferred to a structure known as primary or short-term memory, a process in which attention plays an important role. In the short-term memory, the information is subject to further analysis, whereby the perceptual information is combined with information retrieved from secondary or long-term memory. In the primary memory, processes such as judgment, inference, and problem solving take place. On the basis of an analysis of the meaning of the stimulus input, a response is generated. Finally, a trace of the event is permanently encoded in the secondary memory.

It is in the primary memory that one might assume that the percepts (the perceptual information) are linked with the appropriate emotions, i.e., the emotions that similar circumstances aroused on previous occasions.

The concept of the NA-ergic nature of hedonic disturbances has not been systematically studied and, hence, is largely hypothetical. Some scattered data, admittedly at best tentative in nature, tend to support it. They have been discussed elsewhere (van Praag et al. 1990).

DISCUSSION

A Multiaminergic Hypothesis of Depression

The yield of the research with catecholamine precursors in depression can be measured in scientific rather than in therapeutic terms.

Studies with L-dopa provided support for the hypothesis that diminished dopamine metabolism underlies motor inhibition as it may occur in depressed patients as well as in other psychiatric and nonpsychiatric disease states. In other words, deficient dopaminergic "tone" is not a condition specific for Parkinson's disease, but one specific for motor retardation and lack of initiative, across diagnoses.

Studies with tyrosine derive their significance from the admittedly preliminary observation that tyrosine seems to exert a beneficial effect on the hedonic disturbances as they might occur in depression. L-Dopa failed to exert such an effect. In addition, in humans, tyrosine has a more pronounced effect on NA release than on dopamine release, while the reverse is true for L-dopa. These observations point to NA as possibly involved in the pathogenesis of anhedonia.

A third conclusion drawn from catecholamine precursor studies in depressives is that apparently the increase of central 5-HT availability alone is not a sufficient antidepressant measure and that the best conditions for antidepressant activity are provided by a simultaneous rise of both catecholamines and 5-HT. A substantial body of evidence now suggests that 5-HT disturbances as they might occur in depression are correlated with increased anxiety and disturbed aggression regulation. We have postulated that raising 5-HT availability might ameliorate these components of the depressive syndrome (van Praag et al. 1987b). The data discussed in this chapter permit an addition to this hypothesis, namely that enhancing the availability of dopamine will reduce motor inhibition and enhancing that of NA will improve hedonic receptivity. An optimal antidepressant effect, this hypothesis implies, requires an overall augmentation of monoaminergic functioning (Figure 4-14).

Functional Psychopharmacology

Obviously, dissecting the depressive syndrome into its component parts, i.e., the psychological dysfunctions or psychopathological dimensions, and relating these components to biological dysfunctions has proven to be more fruitful than the classical nosological approach of trying to relate biological variables to disease entities (van Praag et al. 1987c). What I have called the functional approach to psychopathology has important implications, not only for biological psychiatry but for clinical psychopharmacology as well.

Psychotropic drugs, in this case antidepressants, are becoming more and more selective as far as mode of action is concerned, and one may conjecture that this trend will continue. The selective 5-HT uptake inhibitors are a case in point. Through the functional analysis

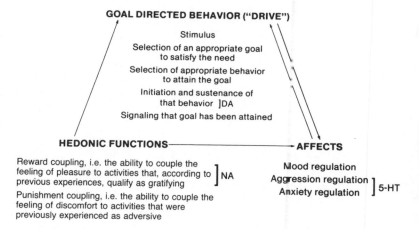

Figure 4-14. Depiction of the major domains of psychological dysfunctioning in major depression, melancholic type, and the hypothesized monoaminergic dysfunctions underlying these psychological dysfunctions.

of biological variables in psychiatric disorders, we have learned that correlations between biological and psychological dysfunctions are much more obvious than those between biological dysfunctions and disease entities. Biochemically selective drugs, therefore, can be expected to exert beneficial effects on particular psychological dysfunctions, irrespective of nosological diagnosis. Since a particular behavior disorder is composed of various psychological dysfunctions, one would often need a combination of (biochemically selective) drugs to combat the multiple psychological dysfunctions—and thus the disease state—adequately. Functional psychopathology, then, will lead to functional psychopharmacology, and functional psychopharmacology will be multipharmaceutical. Medicine provides useful analogies. In the case of a myocardial infarction, for example, treatment is not based on the nosological diagnosis but on its functional sequelae. The cardiac condition is analyzed in terms of, say, disturbed rhythm, regularity, conduction, and many other factors. Treatment is directed toward these dysfunctions and, hence, often several drugs are needed. Goal-directed, dysfunction-oriented pharmacotherapy is *the* scientific approach to drug treatment, at least so long as we cannot address the cause(s) of the diseases directly. It is this direction in which I see clinical psychopharmacology developing or, phrased more cautiously, it is a development that should be actively pursued.

FUTURE OF THE PRECURSOR STRATEGY

Monoamine precursors have provided us with important insights into the relationship between abnormal behavior and the brain. Is their role finished now that more and more synthetic drugs have become available with more and more selective effects on particular aspects of the metabolism of a particular monoamine? I don't think so. Precursors are still the only natural way to force the monoamine system to increase its output. For human brain and behavior studies, this approach still has a lot to offer. Studying the therapeutic usefulness of precursor mixtures in psychiatric disorders seems at this moment to be the most obvious way to go. The field of scientific research is still wide open.

REFERENCES

Ashton H: Brain or Systems Disorders and Psychotropic Drugs. Oxford, UK, Oxford University Press, 1987

Banki CM: Correlation between CSF metabolites and psychomotor activity in affective disorders. J Neurochem 28:255–257, 1977

Banki CM, Molnar G, Vojnik M: Cerebrospinal fluid amine metabolites, tryptophan and clinical parameters in depression. J Affect Disord 3:91–99, 1981

Crow TJ: Catecholamine-containing neurones and electrical self-stimulation: 2. A theoretical interpretation and some psychiatric implications. Psychol Med 3:66–73, 1973

Crow TJ: A general catecholamine hypothesis. Neuroscience Research Progress Bulletin 15:195–205, 1977

Crow TJ, Deakin JFW: Neurohormonal transmission, behaviour and mental disorder, in The Scientific Foundations of Psychiatry. Handbook of Psychiatry, Vol 5. Edited by Shepherd M. Cambridge, UK, Cambridge University Press, 1985, pp 137–182

Franklin KBJ: Catecholamines and self-stimulation: reward and performance effects dissociated. Pharmacol Biochem Behav 9:813–820, 1978

Freed CR, Yamamoto BK: Regional brain dopamine metabolism: a marker for the speed, direction and posture of moving animals. Science 229:62, 1985

Hunt GE, Atrens DM, Chesher GB, et al: α-Noradrenergic modulation of hypothalamic self-stimulation: studies employing clonidine, 1-phenylephrine and α–methyl–p–tyrosine. Eur J Pharmacol 37:105–111, 1976

Kihlstrom JF: The cognitive unconscious. Science 237:1445–1452, 1987

Lakke JPWF, Korf J, van Praag HM, et al: Predictive value of the probenecid test for the effect of L-dopa therapy in Parkinson's disease. Nature 236:208–209, 1972

Lindstrom LH: Low HVA and normal 5-HIAA CSF levels in drug free schizophrenia patients, compared to healthy volunteers: correlations to symptomatology and heredity. Psychiatry Res 14:265–274, 1985

Olds J, Milner P: Positive reinforcement produced by electrical stimulation of the septal area and other regions of the rat brain. J Comp Physiol Psychiatr 47:419–427, 1954

Papeschi R, McClure DJ: Homovanillic and 5-hydroxyindoleacetic acid in cerebrospinal fluid in depressed patients. Arch Gen Psychiatry 25:354–358, 1971

Raleigh MJ: Differential behavioral effects of tryptophan and 5-hydroxytryptophan in vervet monkeys: influence of catecholaminergic systems. Psychopharmacology (Berlin) 93:44–50, 1987

Stein L: Reward transmitters: catecholamines and opioid peptides, in Psychopharmacology: A Generation of Progress. Edited by Lipton MA, DiMascio A, Killam KF. New York, Raven Press, 1978, pp 569–581

van Praag HM: Management of depression with serotonin precursors. Biol Psychiatry 16:291–310, 1981

van Praag HM: In search of the mode of action of antidepressants: 5-HT–tyrosine mixtures in depressions. Neuropharmacology 22:433–440, 1983

van Praag HM: Studies in the mechanism of action of serotonin precursors in depression. Psychopharmacol Bull 20:599–602, 1984

van Praag HM: Serotonin precursors with and without tyrosine in the treatment of depression, in Biological Psychiatry. Edited by Shagrass C, Josias R, Bridger W, et al. New York, Elsevier, 1986, pp 97–99

van Praag HM, Korf J: Retarded depression and the dopamine metabolism. Psychopharmacology (Berlin) 19:199–203, 1971

van Praag HM, Korf J: Central monoamine deficiency in depression: causative or secondary phenomenon? Pharmakopsychiatrie 8:321–326, 1975

van Praag HM, Lemus C: Monoamine precursors in the treatment of psychiatric disorders, in Nutrition and the Brain, Vol 7. Edited by Wurtman RJ, Wurtman JJ. New York, Raven Press, 1986, pp 89–138

van Praag HM, Korf J, Lakke JPWF, et al: Dopamine metabolism in depression, psychoses and Parkinson's disease: the problem of the specificity of biological variables in behaviour disorders. Psychol Med 5:138–146, 1975

van Praag HM, Kahn R, Asnis GM, et al: Denosologization of biological psychiatry on the specificity of 5-HT disturbances in psychiatric disorders. J Affective Disord 13:1–8, 1987a

van Praag HM, Kahn R, Asnis GM, et al: Therapeutic indications for serotonin potentiating compounds: a hypothesis. Biol Psychiatry 22:205–212, 1987b

van Pragg HM, Lemus C, Kahn R: Hormonal probes of central serotonergic activity: do they really exist? Biol Psychiatry 22:86–98, 1987c

van Praag HM, Verhoeven WMA, Kahn RS: Psychofarmaca, 3rd Edition. Van Gorcum, Assen/Maastricht, 1988

van Praag HM, Asnis GM, Brown SL, et al: Beyond serotonin. a multiaminergic perspective on abnormal behavior, in Serotonin in Psychiatric Disorders. Edited by Brown SL, van Praag HM. New York, Brunner/Mazel, 1990

Chapter 5

Plasma Neutral Amino Acids Associated With the Efficacy of Antidepressant Treatment: A Summary

Svend E. Møller, Ph.D.

Chapter 5

Plasma Neutral Amino Acids Associated With the Efficacy of Antidepressant Treatment: A Summary

The general efficacy of antidepressant treatment is unsatisfactorily low. A survey published about 10 years after the appearance of the first tricyclic antidepressant and based on more than a hundred studies revealed that only about two-thirds of depressed inpatients benefit from pharmacotherapeutic treatment (Wechsler et al. 1965). Considerable efforts have been made to develop new compounds with improved efficacy, but the second-generation drugs do not appear to provide significant advantages over the traditional tricyclic antidepressants in terms of clinical efficacy (Kane and Lieberman 1984; Prien and Blaine 1984). At least three major causes may contribute to the variance in treatment outcome.

CLINICAL AND PHARMACOKINETIC VARIABLES

One reason for variance in the therapeutic response is the diagnostic classification. There is evidence that delusional depression represents a subtype of unipolar depression that responds poorly to tricyclic antidepressant therapy (Charney and Nelson 1981; Glassman and Roose 1981). There are also indications that patients with rapid-cycling affective disorders show a relatively poor response to conventional antidepressant medication, as do patients with chronic depression (see Leonard 1988).

Another reason for variance in clinical response to tricyclic therapy is the wide interindividual variations in pharmacokinetic factors that regulate the steady-state serum drug levels. Some antidepressant

This study was supported in part by grants from the Danish Medical Research Council and the Psychiatric Research Foundation.

drugs, e.g., imipramine, may have a lower limit cf serum drug level below which the clinical efficacy is reduced (Glassman et al. 1977; Reisby et al. 1977). Other compounds, like nortriptyline, possibly have a serum level range within which patients receive optimum antidepressant efficacy but above or below which the efficacy is reduced (Åsberg et al. 1971; Kragh-Sørensen et al. 1976).

BIOLOGICAL VARIABLES—MONOAMINE METABOLITES

The third reason for differential response to antidepressant therapy is the patients' biological variation. More than two decades ago, the hypotheses were presented that a functional deficiency of the brain monoamines serotonin (Coppen 1967) and noradrenaline (Schildkraut 1965) may play a principal role in the pathogenesis of depressive disorder. These hypotheses have stimulated a large number of studies of patients with major depression. Particular attention has been paid to the monoamine metabolites 5-hydroxyindoleacetic acid (5-HIAA) and 3-methoxy-4-hydroxyphenyl glycol (MHPG) in biological fluids as possible predictors of treatment response. Some studies have shown greater clinical response in depressive patients with low cerebrospinal fluid (CSF) levels of 5-HIAA to drugs that preferentially act on serotonergic neurons as compared with patients with high levels (Åberg-Wistedt 1981; Bertilsson et al. 1978; van Praag 1977). However, other studies have failed to demonstrate such a relationship (Dahl et al. 1982; Maas et al. 1982; Timmerman et al. 1987).

Measuring the contents of MHPG in 24-hour urine samples resulted initially in promising findings, since it was shown that patients with low MHPG values rather than high levels showed a more favorable response to noradrenaline reuptake inhibitors than to serotonin reuptake inhibitors (Beckmann and Goodwin 1975; Hollister et al. 1980; Maas et al. 1982; Schatzberg et al. 1980). By using the MHPG output level as the criterion for drug selection, a prospective study obtained better clinical results than had been obtained by using more traditional selection methods (Cobbin et al. 1979). Unfortunately, other studies failed to show a relationship between pretreatment urinary MHPG level and the outcome of antidepressant treatment (Coppen et al. 1979; Janicak et al. 1986; Puzynski et al. 1984; Sharma et al. 1986; Spiker et al. 1980).

ENDOCRINOLOGICAL VARIABLES

Neuroendocrinological variables and their relationship to the clinical response in depressed patients to antidepressant treatment have been studied. The outcome of the dexamethasone suppression test was associated with the clinical response to antidepressant therapy in some studies (Brown et al. 1980; Fraser 1983; Rihmer et al. 1985), but not in others (Georgotas et al. 1986; Greden et al. 1981; Larsen et al. 1985; Myers 1988; Sauer et al. 1986). However, there seems to be an agreement that an early dexamethasone suppression test normalization may have predictive value for good treatment response (Albala et al. 1981; Maes et al. 1986; Schweitzer et al. 1987). Also, the endocrinological response to the thyrotropin-releasing hormone test may be useful in the control of continuation therapy with antidepressants in depressed patients rather than as a predictor of therapeutic response (Kirkegaard et al. 1975; Kirkegaard and Smith 1978; Krog-Meyer et al. 1984; Larsen et al. 1985).

PRECURSOR AMINO ACIDS

Another alternative to estimating brain monoaminergic function is determining the availability of the precursor amino acids from plasma to brain. Serotonin is synthesized from and (at a rate that parallels the brain concentration of) its natural precursor tryptophan, and the formation of noradrenaline depends, in part, on the brain tyrosine concentration. Tryptophan and tyrosine are transported from blood to brain by a carrier mechanism specific for those two and the other large neutral amino acids (LNAAs), notably valine, isoleucine, leucine, and phenylalanine (Pardridge 1977; Yuwiler et al. 1977). Because of the competitive nature of the amino acid transport, it is the molar ratio in plasma of (total) tryptophan to the sum of the other LNAAs, rather than the tryptophan concentration alone, that under physiological conditions best predicts the brain tryptophan concentration, and it is the plasma ratio tyrosine/LNAA that predicts the brain tyrosine concentration (Fernstrom and Faller 1978).

CLINICAL TRIALS

In a series of clinical trials performed during the past 10 years and comprising more than 150 inpatients from eight psychiatric centers, I have studied the association between the pretreatment plasma ratios tryptophan/LNAA and tyrosine/LNAA and the therapeutic response in patients with a major depressive episode to a variety of antidepressant treatments.

Patients and Methods

All the patients studied were hospitalized for moderate to severe depression. Some studies included endogenous depressed patients solely; other studies included both endogenous and nonendogenous depressed patients, as classified by means of the Newcastle scales. In terms of psychopathology, the diagnoses endogenous and nonendogenous depression are on the whole concordant with major depressive episode with and without melancholia, respectively (DSM-III; American Psychiatric Association 1980). Patients with severe anxiety or stupor and patients with hallucinations or delusions were excluded from the studies, as were patients with chronic depressive illness (duration more than 1 year) and rapid-cycling affective illness. The studies included both unipolar and bipolar depressed patients, as specified in the original reports.

Further information on the patients, inclusion and exclusion criteria, dosage schedules, serum steady-state drug levels, etc., are given in the original reports. The severity of the depressive state was assessed in all studies by means of the Hamilton Depression Rating Scale (HDRS) items 1 to 17 (Hamilton 1960). The quantitative inclusion criteria were in general a score of at least 18 on the final placebo or washout day. Complete responders had a final score of 7 or less, and partial responders had a score of 8 to 15. Venous blood samples were collected at 8 A.M. from fasting patients, and amino acids in plasma were separated by ion-exchange chromatography and measured spectrophotometrically (Møller 1977).

The relationship between plasma amino acid ratios and clinical improvement was investigated by means of the Spearman rank correlation coefficient (r_s) and one-tailed P value, and the difference in improvement between subgroups of the patients was investigated by means of the Mann-Whitney U-test and one-tailed P value (Siegel 1956). Differences between groups in serum steady-state drug levels were investigated by two-tailed t test.

Results

Imipramine. In a multicenter study, 66 endogenously or nonendogenously depressed patients were allocated to treatment with imipramine for 5 weeks at a fixed dose of 225 mg/day. There was no significant difference in the response rate between the two diagnostic groups, and 68% of the patients showed complete or partial response after 4 weeks' treatment (Reisby et al. 1977).

Before active treatment, venous blood samples were collected from 39 patients with a mean age of 48 years (range 20 to 64 years) from one clinical center to determine plasma amino acids. In the original report on the association between plasma amino acid profiles and clinical response, the plasma tryptophan ratio included only three competing amino acids in the denominator, and the slight influence of age on the amino acid ratios was accounted for by means of regression technique (Møller 1985; Møller et al. 1981). The data have now been reevaluated using the conventional plasma tryptophan ratio with five amino acids in the denominator and without age correction. The sum of the pretreatment plasma ratios tryptophan/LNAA and tyrosine/LNAA correlated significantly positively with the final HDRS score ($r_s = 0.36$, $P = .012$) and significantly inversely with the percentage of HDRS score ($r_s = -0.31$, $P = .027$; Figure 5-1).

For imipramine, several studies have found evidence that the clinical response is more or less associated with serum drug levels (e.g., Glassman et al. 1977; Reisby et al. 1977). In the present patient sample, the steady-state serum concentration of imipramine plus desipramine correlated inversely with the final HDRS score ($r_s = -0.29$, $P = .036$) and directly with the percentage reduction of the score ($r_s = 0.30$, $P = .035$). Furthermore, there are indications that a steady-state ratio of imipramine to desipramine in serum below 0.2 is associated with less improvement (Møller et al. 1985a). Obviously, it is necessary to consider both the amino acid profile and the serum drug level.

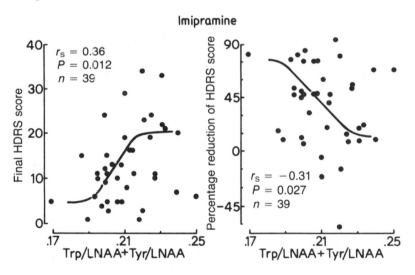

Figure 5-1. Relation between the sum of the pretreatment plasma ratios tryptophan/LNAA and tyrosine/LNAA and the final Hamilton Depression Rating Scale (HDRS) score (left) and percentage reduction of HDRS score (right) of 39 depressed patients treated with imipramine for 5 weeks. r_s = Spearman rank correlation coefficient. Trp = tryptophan. Tyr = tyrosine.

The patient sample was divided into a half part with a sum of plasma ratios tryptophan/LNAA and tyrosine/LNAA below 0.210 ($n = 19$) and a half part with a sum of the ratios above that limit ($n = 20$). Although there was no difference between the two groups in the steady-state serum imipramine plus desipramine levels, the patients with a sum of plasma tryptophan and tyrosine ratios below the mean showed significantly greater reduction in score than the other patients ($P = .01$, U-test). Second, the patients in each half group were separated according to a previously identified serum level limit of imipramine plus desipramine of 180 ng/ml (Glassman et al. 1977; Møller et al. 1985a). The amelioration curves of the patients in the four subgroups are shown in Figure 5-2. Patients with a sum of

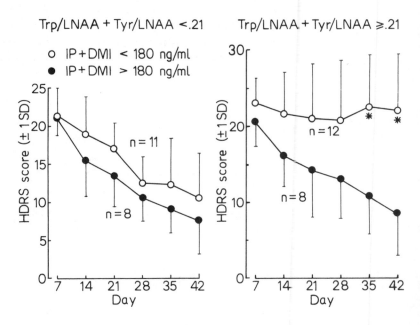

Figure 5-2. Mean ± 1 SD HDRS scores as a function of treatment time with imipramine of 19 depressed patients with a sum of pre-treatment plasma ratios tryptophan/LNAA and tyrosine/LNAA below 0.210 (left) and of 20 depressed patients with a sum of these ratios above 0.210 (right). Day 7 is final placebo day. Open and filled circles indicate a serum steady-state level of imipramine plus desipramine of less and more than 180 ng/ml, respectively. Asterisks indicate significant difference in improvement ($P = .002$, U-test).

tryptophan and tyrosine ratios below 0.210 showed nearly the same improvement over time whether their steady-state serum concentration of imipramine plus desipramine was below or above 180 ng/ml. Patients with a sum of ratios tryptophan/LNAA and tyrosine/LNAA above that limit differed markedly in clinical response; the patients with a serum imipramine plus desipramine level above 180 ng/ml showed nearly the same amelioration curve as the former two groups, whereas the patients with a serum drug level below 180 ng/ml produced a virtual placebo response curve and showed significantly smaller improvement after 4 and 5 weeks of treatment than the patients with the comparable amino acid profile with a higher serum drug level ($P = .002$, U-test).

Obviously, only 50% of the whole patient sample received optimal individual therapy on the applied schematic dosage schedule. Eight patients with tryptophan/LNAA plus tyrosine/LNAA below 0.210 might have shown the same therapeutic response with a serum imipramine plus desipramine level below 180 ng/ml, and 12 patients with a sum of tryptophan and tyrosine ratios above that limit had serum drug levels too low to benefit from treatment.

Amitriptyline. Twenty-one endogenously depressed patients with a mean age of 57 years (range 34 to 75 years) were treated for 4 weeks with amitriptyline in a sustained release preparation in a dose, in most cases, of 100 mg/day (Honoré et al. 1982). The applied dose is equivalent to 150 mg/day of conventional amitriptyline in terms of clinical efficacy (Liisberg et al. 1978; Sedman 1973). After 4 weeks' treatment, 71% of the patients showed complete or partial response. Since the plasma trytophan ratio was originally reported with three amino acids in the denominator (Møller 1985; Møller et al. 1983a), the data have been reevaluated, and Figure 5-3 shows the conventional pretreatment plasma ratio tryptophan/LNAA correlated positively with the final HDRS score ($r_S = 0.44$, $P = .024$) and inversely with the percentage reduction of the score ($r_S = -0.43$, $P = .027$) after 4 weeks' treatment. There was no significant correlation between the plasma ratio tyrosine/LNAA and the final score ($r_S = 0.21$, NS) or the percentage reduction of the score ($r_S = -0.19$, NS).

In the three major trials of imipramine, amitriptyline, and nortriptyline, each including more than 20 patients, the relationship between pretreatment plasma amino acid ratio and clinical response most likely appeared sigmoidally. This is recognized after analysis of the data by the moving average technique, i.e., by calculating all the means of five successive data points. As shown in Figure 5-4, which is based on exactly the same data as Figure 5-3, this technique greatly minimizes the variance so that the "true" amino acid ratio to clinical response relationship emerges.

When the patient sample was divided in two half parts by the mean of the plasma ratio tryptophan/LNAA (0.090), the 10 patients below that limit showed significantly greater improvement ($P = .009$, U-test) than the 11 patients above that limit with comparable serum steady-state drug levels (Table 5-1).

Nortriptyline. Twenty-six endogenously depressed patients with a mean age of 53 years (range 28 to 73 years) were treated for 4 weeks with nortriptyline in doses adequate to achieve a steady-state serum nortriptyline concentration between 70 and 130 ng/ml. After 4 weeks' treatment, 50% of the patients showed complete or partial response (Møller et al. 1985b).

The pretreatment plasma ratio tyrosine/LNAA was positively correlated with the final HDRS score ($r_s = 0.51$, $P = .004$) and inversely with the percentage reduction of the score ($r_s = -0.53$, $P = .003$; Figure 5-5). There was no significant correlation between the pretreatment plasma ratio tryptophan/LNAA and the final score ($r_s = 0.07$, NS) or percentage reduction of the score ($r_s = -0.09$, NS),

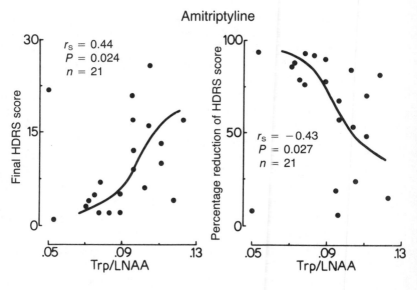

Figure 5-3.   Relation between the pretreatment plasma ratio tryptophan/LNAA and the final HDRS score (left) and percentage reduction of HDRS score (right) of 21 depressed patients treated with amitriptyline for 4 weeks.

whereas the pretreatment plasma ratio phenylalanine/LNAA correlated slightly with the two clinical variables (r_s = 0.32, P = .056, and r_s = −0.26, P = .10, respectively).

To evaluate the influence of the serum nortriptyline level on the variance in therapeutic response, the patient sample was divided by its mean of the plasma ratio tyrosine/LNAA (0.121). The 12 patients with ratio tyrosine/LNAA below that limit showed significantly greater improvement than the 14 patients above that limit with comparable serum nortriptyline concentrations (P = .009, U-test; Table 5-2).

Citalopram. The selective serotonin uptake inhibitor citalopram was administered in a dose generally of 60 mg/day for 4 weeks to eight endogenously and six nonendogenously depressed patients with a mean age of 51 years (range 33 to 64 years) (Timmerman et al. 1987).

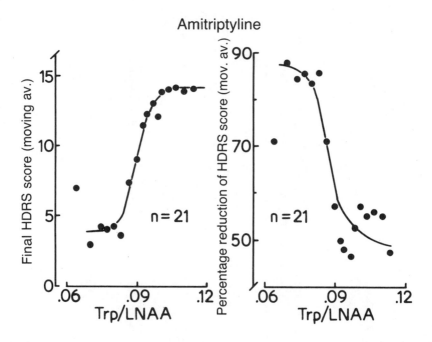

Figure 5-4. Relation between the pretreatment plasma ratio tryptophan/LNAA and the final HDRS score (left) and percentage reduction of HDRS score (right) of 21 depressed patients treated with amitriptyline for 4 weeks illustrated by means of the moving average technique.

Table 5-1. Hamilton Depression Rating Scale scores and steady-state serum drug levels of depressed patients treated with amitriptyline

Ratio tryptophan/LNAA	n	HDRS score		Percentage reduction of score	Serum concentration (ng/ml)		
		Day 0	Day 28		Ami	Nor	Ami + Nor
Below 0.090	10	25 ± 4	5 ± 6	78 ± 26	82 ± 76	54 ± 29	136 ± 98
Above 0.090	11	26 ± 6	14 ± 7	48 ± 28	93 ± 46	83 ± 42	175 ± 84
P		NS	= .009*	= .005*	NS	NS	NS

Note. Results are presented as the mean ± 1 SD. Ami = amitriptyline. Nor = nortriptyline.
*By the *U*-test.

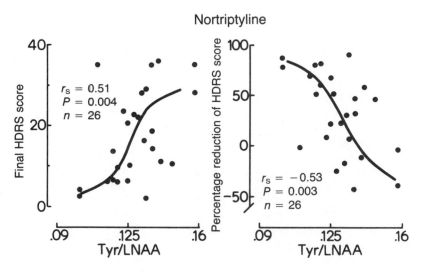

Figure 5-5. Relation between the pretreatment plasma ratio tyrosine/LNAA and the final HDRS score (left) and percentage reduction of HDRS score (right) of 26 depressed patients treated with nortriptyline for 4 weeks.

Table 5-2. Hamilton Depression Rating Scale scores and steady-state serum drug levels of depressed patients treated with nortriptyline

Ratio tyrosine/ LNAA	n	HDRS score Day 0	HDRS score Day 28	Percentage reduction of score	Serum nortriptyline (ng/ml)
Below 0.121	12	25 ± 6	12 ± 10	55 ± 30	95 ± 15
Above 0.121	14	25 ± 4	22 ± 10	14 ± 39	105 ± 16
P		NS	= .009*	= .007*	NS

Note. Results are presented as the mean ± 1 SD.
*By the *U*-test.
Source. Adapted from Møller et al. 1985a, with permission from Elsevier Science Publishers B.V. (Biomedical Division). Copyright 1985.

There was no significant difference between the two diagnostic groups in improvement ($P = .13$, U-test), and 57% of the patients showed complete or partial response.

The pretreatment plasma ratio tryptophan/LNAA correlated significantly with the final HDRS score ($r_s = 0.51$, $P = .030$) as did the plasma ratio tyrosine/LNAA ($r_s = 0.56$, $P = .018$), whereas the two amino acid ratios correlated only slightly with the percentage reduction of the score ($r_s = -0.41$, $P = .075$, and $r_s = -0.45$, $P = .052$, respectively). While the ratios tryptophan/LNAA and tyrosine/LNAA were not mutually correlated, the sum of the two plasma amino acid ratios was significantly correlated with the final HDRS score ($r_s = 0.57$, $P = .018$), and the percentage reduction of the score ($r_s = -0.48$, $P = .042$; Figure 5-6 [Møller et al. 1986]).

There was a trend toward greater improvement in patients with a sum of plasma ratios tryptophan/LNAA and tyrosine/LNAA below the mean of the full sample (0.184) than in patients with a sum of ratios above that limit ($P = .054$, U-test) with comparable steady-state serum citalopram concentrations (Table 5-3).

Maprotiline. The selective noradrenaline uptake inhibitor maprotiline was administered in a dose generally of 150 mg/day for 4 weeks to eight endogenously and five nonendogenously depressed

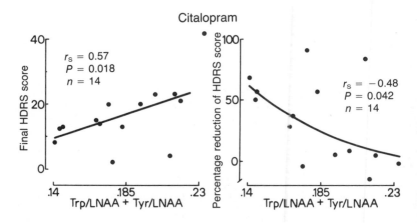

Figure 5-6.   Relation between the sum of the pretreatment plasma ratios tryptophan/LNAA and tyrosine/LNAA and the final HDRS score (left) and percentage reduction of HDRS score (right) of 14 depressed patients treated with citalopram for 4 weeks. Adapted from Møller et al. 1986, with permission from Springer-Verlag. Copyright 1986.

patients with a mean age of 56 years (range 41 to 64 years) (Timmerman et al. 1987). The endogenously depressed patients showed significantly greater clinical improvement than the nonendogenous depressive patients (P = .009, U-test). Sixty-two percent of the patients showed complete or partial response.

The plasma ratio tryptophan/LNAA showed no significant correlation with the final HDRS score (r_s = 0.37, NS) or the percentage reduction of the score (r_s = −0.27, NS), whereas the pretreatment plasma ratio tyrosine/LNAA was significantly correlated with the two specified clinical variables (r_s = −0.53, P = .032, and r_s = 0.59, P = .017, respectively; Figure 5-7). The reason the amino acid ratio to response relationship for maprotiline is inverse to those observed in the other studies is not clear. One possible explanation is that the correlation between the ratio tyrosine/LNAA and the percentage reduction of the HDRS score is almost exclusively attributed to the nonendogenous group (r_s = 0.9, P = .042, versus r_s = 0.19, NS, in the endogenous group). When the eight endogenously depressed patients were considered alone, the optimum relationships were found between the plasma ratio tryptophan/LNAA and the final HDRS score (r_s = 0.53, P = .098) and the percentage reduction of the score (r_s = −0.50, P > .1 [Møller et al. 1986]).

The seven patients with plasma ratios tyrosine/LNAA above the mean of the total sample (0.106) showed significantly greater clinical improvement than the six patients with a lower plasma tyrosine ratio with comparable steady-state serum maprotiline concentrations (Table 5-4).

Electroconvulsive Therapy. Fourteen endogenously depressed patients with a mean age of 61 years (range 43 to 83 years) were treated with unilateral electroconvulsive therapy (ECT) twice weekly.

Table 5-3. Hamilton Depression Rating Scale scores and steady-state serum drug levels of depressed patients treated with citalopram

Ratios tryptophan/ LNAA + tyrosine/LNAA	*n*	HDRS score Day 0	HDRS score Day 28	Percentage reduction of score	Serum citalopram (nmol/l)
Below 0.184	8	25 ± 4	12 ± 6	48 ± 29	132 ± 56
Above 0.184	6	26 ± 8	22 ± 12	14 ± 35	135 ± 52
P		NS	= .054*	= .054*	NS

Note. Results are presented as the mean ± 1 SD.
*By the U-test.

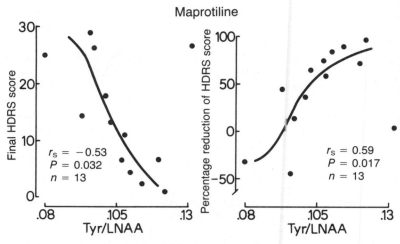

Figure 5-7.   Relation between the pretreatment plasma ratio tyrosine/LNAA and the final HDRS score (left) and percentage reduction of HDRS score (right) of 13 depressed patients treated with maprotiline for 4 weeks. Adapted from Møller et al. 1986, with permission from Springer-Verlag. Copyright 1986.

Treatment was withdrawn when the patients had either recovered or were refractory after 10 to 13 treatments, or when complications arose. Seventy-one percent of the patients showed complete or partial response. All patients remained in the study for at least 2 weeks; data were therefore evaluated on day 14 after four ECTs (Møller and Fogh 1988).

Table 5-4.   Hamilton Depression Rating Scale scores and steady-state serum drug levels of depressed patients treated with maprotiline

Ratio tyrosine/ LNAA	n	HDRS score Day 0	HDRS score Day 28	Percentage reduction of score	Serum maprotiline (nmol/l)
Below 0.106	6	26 ± 6	21 ± 7	13 ± 43	64 ± 34
Above 0.106	7	25 ± 4	8 ± 9	68 ± 31	67 ± 37
P		NS	= .051*	= .011*	NS

Note.   Results are presented as the mean ± 1 SD.
*By the *U*-test.

In the total patient sample, there was no trend or significant relationship between plasma amino acid ratios and clinical improvement. However, for unknown reasons, the 10 female depressed patients showed significantly greater reduction and percentage reduction of HDRS score on day 14 then the 4 males ($P = .038$, U-test). When the 10 females were considered alone, the plasma ratio tryptophan/LNAA showed no significant correlation with the HDRS score ($r_s = 0.19$, NS) or the percentage reduction of the score ($r_s = -0.11$, NS) on day 14, whereas the pretreatment plasma ratio tyrosine/LNAA correlated with these clinical variables ($r_s = 0.58$, $P = .044$, and $r_s = -0.47$, $P = .082$, respectively). A bit stronger correlation was observed between the sum of plasma ratios tryptophan/LNAA and tyrosine/LNAA and the HDRS score ($r_s = 0.64$, $P = .027$) and the percentage reduction of the score on day 14 ($r_s = -0.51$, $P = .066$; Figure 5-8).

The 10 female depressed patients had a mean of the sum of ratios tryptophan/LNAA and tyrosine/LNAA of 0.193. The six female patients with a sum of amino acid ratios below that limit showed a trend towards greater reduction of the HDRS score ($P = .072$, U-test) and a significantly greater percentage reduction of the score on day 14 ($P = .033$, U-test) than the female depressed patients above that limit with comparable seizure durations (Table 5-5).

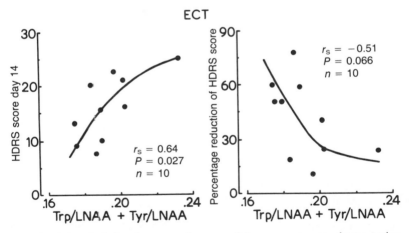

Figure 5-8. Relation between the sum of the pretreatment plasma ratios tryptophan/LNAA and tyrosine/LNAA and the final HDRS score (left) and percentage reduction of HDRS score (right) of 10 female depressed patients treated with four ECTs during 2 weeks.

Clomipramine. Clomipramine was administered as the one drug in a comparative multicenter trial to 46 endogerously or nonendogenously depressed patients for at least 4 weeks in a fixed dose of 150 mg/day (Danish University Antidepressant Group, in press). Seventeen of these patients with a mean age of 49 years (range 22 to 67 years) from four clinical centers were included in the amino acid study, and all showed complete or partial response after 4 weeks' treatment.

There was no trend or significant correlation between the ratio tyrosine/LNAA and clinical improvement, whereas the pretreatment plasma ratio tryptophan/LNAA was inversely correlated with the reduction of HDRS score ($r_s = -0.51$, $P = .020$) and the percentage reduction of the score ($r_s = -0.47$, $P = .029$; Figure 5-9). The failure of the ratio tryptophan/LNAA to correlate significantly with the final HDRS score ($r_s = 0.27$, NS) may be explained by a significant inverse correlation between the tryptophan ratio and the starting score in this particular patient sample ($r_s = -0.45$, $P = .038$), which may have partly counterbalanced the anticipated association between the tryptophan ratio and the final score (Møller et al. 1989). In a previous trial of clomipramine on depressed inpatients, there was a trend toward a positive correlation between a modified pretreatment plasma tryptophan ratio and the final HDRS score (Møller 1985).

The nine patients with plasma ratio tryptophan/LNAA below the mean of the full sample (0.080) showed significantly greater reduction of HDRS score ($P = .033$, U-test) and a trend toward greater percentage reduction of the score ($P = .067$, U-test) than the eight

Table 5-5.   Hamilton Depression Rating Scale scores and seizure durations for female depressed patients treated with ECT

Ratios tryptophan/ LNAA + tyrosine/LNAA	n	HDRS score		Percentage reduction of score	Seizure duration (sec)
		Day 0	Day 14		
Below 0.193	6	27 ± 6	13 ± 5	52 = 18	47 ± 11
Above 0.193	4	28 ± 7	21 ± 4	21 = 4	50 ± 21
P		NS	= .072*	= .033*	NS

Note.   Results are presented as the mean ± 1 SD. Each patient received four ECTs. Mean and SD for seizure duration are based on 24 and 16 treatments, respectively.
*By the U-test.

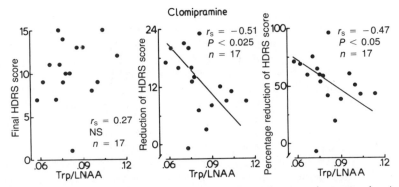

Figure 5-9. Relation between the pretreatment plasma ratio tryptophan/ LNAA and the final HDRS score (left), the reduction of HDRS score (middle), and the percentage reduction of HDRS score (right) of 17 depressed patients treated with clomipramine for 4 weeks. Reprinted from Møller et al. 1990, with permission of Elsevier Science Publishing Co., Inc. Copyright 1990.

patients with a higher ratio tryptophan/LNAA after 4 weeks' treatment (Table 5-6).

Paroxetine. The selective serotonin uptake inhibitor paroxetine was administered as the comparison drug in the above multicenter study to 56 endogenously or nonendogenously depressed patients in a fixed

Table 5-6. Hamilton Depression Rating Scale scores and steady-state serum drug levels of depressed patients treated with clomipramine

Ratio tryptophan/ LNAA	n	HDRS score		Percentage reduction of score	Serum clomipramine (ng/ml)
		Day 7	Day 35		
Below 0.080	9	25 ± 5	10 ± 3	56 ± 25	112 ± 57
Above 0.080	8	20 ± 3	10 ± 4	49 ± 22	101 ± 50
P		= .033**	.033*	.067*	NS**

Note. Results are presented as the mean ± 1 SD. Day 7 is final placebo day.
*By the U-test. **By the two-tailed t test.
Source. Reprinted from Møller et al. 1990, with permission from Elsevier Science Publishing Co., Inc. Copyright 1990.

dose of 30 mg/day for at least 4 weeks (Danish University Antidepressant Group, in press). Twenty-seven of these patients with a mean age of 50 years (range 32 to 67 years) from four clinical centers were included in the amino acid study, and 70% of the patients showed complete or partial response after 4 weeks' treatment.

There was no trend toward or significant correlation between the plasma tryptophan/LNAA or tyrosine/LNAA and the clinical improvement in the sample of 27 depressive patients. Because of these unexpected results on the amino acid ratio to response relationship, the data for 15 patients from one clinical center were analyzed separately from the data of the remaining 12 patients from the other three clinical centers. In the former group, there was no trend or significant relationship between the plasma ratio tryptophan/LNAA and the final HDRS score ($r_s = 0.21$, NS), the reduction of the score ($r_s = -0.03$, NS), or the percentage reduction of the score ($r_s = 0.09$, NS). In the latter group of 12 patients, the pretreatment plasma ratio tryptophan/LNAA was significantly correlated with the final HDRS score ($r_s = 0.60$, $P = .021$), the reduction of the score ($r_s = -0.52$, $P = .043$), and the percentage reduction of the score ($r_s = -0.49$, $P = .051$; Figure 5-10), whereas the plasma ratio tyrosine/LNAA showed no trend or significant correlation with the specified clinical variables (Møller et al. 1989). At present, there is no ready explanation for the difference between the two patient groups in the relationship of amino acid ratio to clinical response.

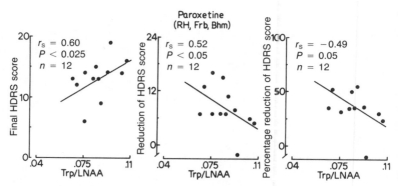

Figure 5-10.   Relation between the pretreatment plasma ratio tryptophan/LNAA and the final HDRS score (left), reduction of HDRS score (middle), and percentage reduction of HDRS score (right) of 12 depressed patients treated with paroxetine for 4 weeks. Reprinted from Møller et al 1990, with permission of Elsevier Science Publishing Co., Inc. Copyright 1990.

There was a trend toward a greater reduction of HDRS score in patients with plasma ratios tryptophan/LNAA below the mean of the total sample (0.086) than in patients with a higher ratio. The difference between the subgroups in percentage reduction of the score was modest (Table 5-7).

Summary and Conclusions

The relationship between the pretreatment plasma ratios tryptophan/LNAA and tyrosine/LNAA on the one side and the final HDRS score, the reduction, and the percentage reduction of the score on the other side for the various trials of antidepressant treatments is summarized in Table 5-8. The results strongly suggest that biological variables in patients suffering from a major depressive episode determine in part the clinical response to antidepressant therapy. However, the curves for the relationship of the amino acid ratio to response indicate that rather than biochemical subgroups within patient populations, the patients seem to be confined to a biochemical continuum in which there is a gradual change of treatment efficacy from the one end to the other.

On average, the correlation coefficients were close to 0.5, indicating that about 25% of the variance in response is associated with the plasma ratio tryptophan/LNAA or tyrosine/LNAA or their sum. Because of the relatively large remainder variance, it is not possible to predict response in individual depressed patients. On the other hand, the results suggest that the general efficacy of antidepressant treatment in populations of depressed patients can be improved by select-

Table 5-7. Hamilton Depression Rating Scale scores and steady-state serum drug levels of depressed patients treated with paroxetine

Ratio tryptophan/ LNAA	n	HDRS score Day 7	HDRS score Day 35	Percentage reduction of score	Serum paroxetine (ng/ml)
Below 0.086	6	23 ± 4	12 ± 3	46 ± 15	86 ± 43
Above 0.086	6	20 ± 2	14 ± 3	28 ± 22	67 ± 46
P		NS	= .057*	= .12*	NS

Note. Results are presented as the mean ± 1 SD. Day 7 is final placebo day.
*By the *U*-test.

Table 5-8. Correlation coefficients (r_S) and P values for the relation between pretreatment plasma amino acid ratios and clinical improvement in trials of various antidepressant treatments

Treatment	Duration (weeks)	n	Ratio	Final HDRS score		Reduction of score		Percentage reduction of score	
				r_S	P	r_S	P	r_S	P
Imipramine	5	39	tryptophan/LNAA + tyrosine/LNAA	.36	.012	−.31	.027	−.31	.027
Amitriptyline	4	21	tryptophan/LNAA	.44	.024	−.31	.087	−.43	.027
Nortriptyline	4	26	tyrosine/LNAA	.51	.004	−.47	.008	−.53	.003
Citalopram	4	14	tryptophan/LNAA + tyrosine/LNAA	.57	.018	−.46	.051	−.48	.042
Maprotiline	4	13	tyrosine/LNAA	−.53	.032	.49	.044	.59	.017
ECT	2	10	tryptophan/LNAA + tyrosine/LNAA	.64	.027	−.44	.098	−.51	.066
Clomipramine	1	17	tryptophan/LNAA	.27	.15	−.51	.020	−.47	.029
Paroxetine	4	12	tryptophan/LNAA	.60	.021	−.52	.043	.49	.051

ing the patients according to plasma amino acid ratios. This is illustrated by the three large trials on imipramine, amitriptyline, and nortriptyline, each including more than 20 patients. The percentage of patients with no, mild, and moderate to severe depression in relation to treatment duration with and without selection by plasma amino acid ratios is shown in Figure 5-11.

In the total patient sample consisting of 86 depressed patients, 30% showed complete response, whereas 40% suffered from a moderate to severe depression after treatment for 4 weeks. When this sample is divided into two according to plasma amino acid ratios, the one group with 41 patients ended up with 49% of complete responders and 17% of nonresponders. The efficacy of the treatment in this subgroup is significantly greater than that in the total sample as estimated by the

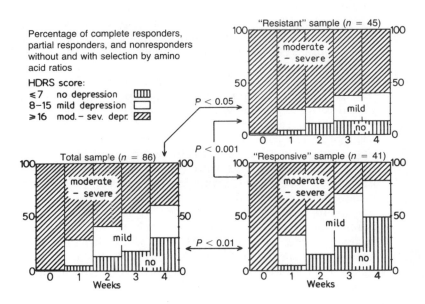

Figure 5-11. Clinical improvement as a function of treatment duration in 86 depressed patients treated with imipramine, amitriptyline, or nortriptyline (left, bottom), in the half group of 41 depressed patients who had pretreatment plasma amino acid ratios below the means of the full patient sample (right, bottom), and in the half group of 45 patients with plasma amino acid ratios above those limits (right, top). The difference between groups in percentage reduction of HDRS score at 4 weeks was estimated by the two-tailed t test.

percentage reduction of HDRS score at 4 weeks ($P = .0098$ by two-tailed t test). The figure suggests that, other things being equal, the response rate might increase further in this "responsive" sample by prolonging the treatment duration beyond 4 weeks.

The other group, consisting of 45 patients, ended up with 60% of patients with moderate to severe depression and 13% with no depression after 4 weeks' treatment. The efficacy is significantly smaller than in the total sample ($P = .039$ by two-tailed t test) and that in the other group ($P > .001$ by two-tailed t test). The figure does not seem to indicate that prolonged treatment with an unchanged treatment schedule would increase the response rate significantly in the "resistant" patient sample. There was no difference between the "responsive" and the "resistant" group in age distribution ($P = .32$ by the one-tailed Kolmogorov-Smirnov two-sample test). Because the patient groups "resistant" to treatment with imipramine, amitriptyline, and nortriptyline are not congruent in regard to the pretreatment plasma amino acid ratios, only about 30% of a patient population can be considered as refractory to treatment with either of these drugs (Figure 5-12).

Other plasma amino acid variables have been investigated as possible predictors of treatment response. Tryptophan circulates in the plasma with a minor fraction free and a major fraction loosely bound to albumin, and suggestions have been made that only the free plasma tryptophan can gain access to the brain. However, there was no significant relationship between the pretreatment free plasma tryptophan level in patients with endogenous depression and the therapeutic response to L-tryptophan (Møller et al. 1979) or amitriptyline (Møller et al. 1983a). Even though the free plasma tryptophan was significantly positively correlated with the total (free plus albumin-bound) plasma tryptophan in healthy controls and endogenously depressed patients (Møller et al. 1983b), the plasma ratio free tryptophan/LNAA was inferior to the ratio (total) tryptophan/LNAA in regard to correlation with the antidepressant response to amitriptyline (Møller 1988).

Another factor that could possibly influence the power of the association between the plasma amino acid ratios and clinical improvement is confined to the carrier-mediated transport of the LNAA across the blood-brain barrier. Each LNAA has a different affinity for the neutral amino acid transport system (Pardridge 1977), which means that some of the LNAAs are more powerful competitive inhibitors of tryptophan transport than others. The plasma amino acid ratios have been recalculated in regard to the difference in affinity of the individual LNAA for the carrier. The regular plasma ratios tryp-

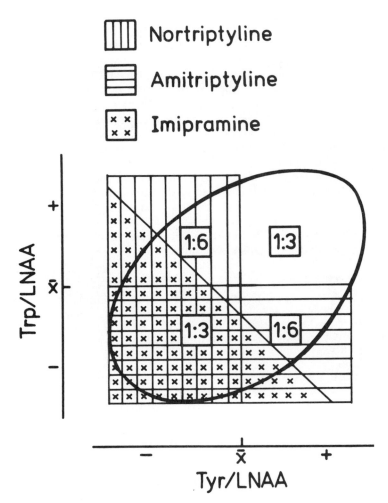

Figure 5-12. Relationship between the pretreatment plasma ratios tryptophan/LNAA and tyrosine/LNAA in depressed patients and the antidepressive response to nortriptyline, amitriptyline, and imipramine. Oval includes majority of depressed patients. Hatched areas indicate biochemical half groups of depressed patients who showed significantly greater clinical response to the antidepressants concerned than the respective complementary groups with comparable mean serum steady-state drug levels. Reprinted from Møller and Kirk 1987, with permission from Elsevier Science Publishing Co., Inc. Copyright 1987.

tophan/LNAA and tyrosine/LNAA cited in this chapter were more strongly associated with the clinical response in depressed patients to antidepressive therapy than were the respective normalized ratios in which the difference in affinity of each LNAA for the transport carrier at the blood-brain barrier has been taken into account (Møller 1988).

In conclusion, the pretreatment plasma ratios tryptophan/LNAA and tyrosine/LNAA correlated significantly with the clinical improvement in hospitalized patients with a major depressive episode to a variety of antidepressant treatments. Although this relationship does not allow prediction of response in individual patients, the clinical improvement in half groups of samples of depressed patients characterized by the plasma amino acid profile was generally significantly greater than in the complementary half groups. The findings suggest that the efficacy of antidepressant treatment can be improved following selection of depressed patients according to the plasma ratios tryptophan/LNAA and tyrosine/LNAA.

REFERENCES

Åberg-Wistedt A, Jostell K-G, Ross SB, et al: Effects of zimelidine and desipramine on serotonin and noradrenaline uptake mechanisms in relation to plasma concentrations and to therapeutic effects during treatment of depression. Psychopharmacology (Berlin) 74:297–305, 1981

Albala AA, Greden JF, Tarika J, et al: Changes in serial dexamethasone suppression tests among unipolar depressives receiving electroconvulsive treatment. Biol Psychiatry 16:551–560, 1981

American Psychiatric Association: Diagnostic and Statistical Manual of Mental Disorders (Third Edition). Washington, DC, American Psychiatric Association, 1980

Åsberg M, Cronholm B, Sjöqvist F, et al: Relationship between plasma level and therapeutic effect of nortriptyline. Br Med J 3:331–334, 1971

Beckmann H, Goodwin FK: Antidepressant response to tricyclics and urinary MHPG in unipolar patients. Arch Gen Psychiatry 32:17–21, 1975

Bertilsson L, Åsberg M, Mellström B, et al: Factors determining drug effects in depressed patients—studies of nortriptyline and chlorimipramine, in Depressive Disorders. Edited by Garattini S. Stuttgart, F.K. Schattauer Verlag, 1978, pp 281–292

Brown WA, Haier RJ, Qualls CB: Dexamethasone suppression test identifies subtypes of depression which respond to different antidepressants. Lancet 1:928–929, 1980

Charney DS, Nelson JC: Delusional and nondelusional unipolar depression: further evidence for distinct subtypes. Am J Psychiatry 138:328–333, 1981

Cobbin DM, Requin-Blow B, Williams LR, et al: Urinary MHPG levels and tricyclic antidepressant drug selection. Arch Gen Psychiatry 36:1111–1115, 1979

Coppen A: The biochemistry of affective disorders. Br J Psychiatry 113:1237–1264, 1967

Coppen A, Rao VAR, Ruthven CRJ, et al: Urinary 4-hydroxy-3-methoxyphenylglycol is not a predictor for clinical response to amitriptyline in depressive illness. Psychopharmacology (Berlin) 64:95–97, 1979

Dahl L-E, Lundin L, le Fèvre Honoré P, et al: Antidepressant effect of femoxetine and desipramine and relationship to the concentration of amine metabolites in cerebrospinal fluid. Acta Psychiatr Scand 66:9–17, 1982

Danish University Antidepressant Group: Paroxetine: a selective serotonin reuptake inhibitor showing better tolerance, but weaker antidepressant effect than clomipramine in a controlled multi-center study. J Affective Disord (in press)

Fernstrom JD, Faller DV: Neutral amino acids in the brain: changes in response to food ingestion. J Neurochem 30:1531–1538, 1978

Fraser AR: Choice of antidepressant based on the dexamethasone suppression test. Am J Psychiatry 140:786–787, 1983

Georgotas A, Stokes P, McCue RE, et al: The usefulness of DST in predicting response to antidepressants: a placebo-controlled study. J Affective Disord 11:21–28, 1986

Glassman AH, Roose SP: Delusional depression. Arch Gen Psychiatry 38:424–427, 1981

Glassman AH, Perel JM, Shostak M, et al: Clinical implications of imipramine plasma levels for depressive illness. Arch Gen Psychiatry 34:197–204, 1977

Greden JF, Kronfol Z, Gardner R, et al: Dexamethasone suppression test and selection of antidepressant medications. J Affective Disord 3:389–396, 1981

Hamilton M: A rating scale for depression. J Neurol Neurosurg Psychiatry 23:56–62, 1960

Hollister LE, Davis KL, Berger PA: Subtypes of depression based on excretion of MHPG and response to nortriptyline. Arch Gen Psychiatry 37:1107–1110, 1980

Honoré P, Møller SE, Jørgensen A: Lithium + L-tryptophan compared with amitriptyline in endogenous depression. J Affective Disord 4:79–82, 1982

Janicak PG, Davis JM, Chan C, et al: Failure of urinary MHPG levels to predict treatment response in patients with unipolar depression. Am J Psychiatry 143:1398–1402, 1986

Kane JM, Lieberman J: The efficacy of amoxapine maprotiline, and trazodone in comparison to imipramine and amitriptyline: a review of the literature. Psychopharmacol Bull 20:240–249, 1984

Kirkegaard C, Smith E: Continuation therapy in endogenous depression controlled by changes in the TRH stimulation test. Psychol Med 8:501–503, 1978

Kirkegaard C, Nørlem N, Lauridsen UB, et al: Prognostic value of thyrotropin-releasing hormone stimulation test in endogenous depression. Acta Psychiatr Scand 52:170–177, 1975

Kragh-Sørensen P, Hansen CE, Baastrup PC, et al: Self-inhibiting action of nortriptyline's antidepressive effect at high plasma levels. Psychopharmacologia 45:305–312, 1976

Krog-Meyer J, Kirkegaard C, Kijne B, et al: Prediction of relapse with the TRH test and prophylactic amitriptyline in 39 patients with endogenous depression. Am J Psychiatry 141:945–948, 1984

Larsen JK, Bjørum N, Kirkegaard C, et al: Dexamethasone suppression test, TRH test and Newcastle II depression rating in the diagnosis of depressive disorders. Acta Psychiatr Scand 71:499–505, 1985

Leonard BE: Biochemical aspects of therapy-resistant depression. Br J Psychiatry 152:453–459, 1988

Liisberg P, Mose H, Amdisen A, et al: A clinical trial comparing sustained release amitriptyline (Saroten® Retard) and conventional amitriptyline tablets (Saroten®) in endogenously depressed patients with simultaneous determination of serum levels of amitriptyline and nortriptyline. Acta Psychiatr Scand 57:426–435, 1978

Maas JW, Kocsis JH, Bowden CL, et al: Pre-treatment neurotransmitter metabolites and response to imipramine or amitriptyline treatment. Psychol Med 12:37–43, 1982

Maes M, De Ruyter M, Hobin P, et al: Repeated dexamethasone suppression test in depressed patients. J Affective Disord 11:155–172, 1986

Møller SE: Quantitative determination of branched-chain and aromatic amino acids by ion-exchange chromatography. Anal Biochem 79:590–593, 1977

Møller SE: Tryptophan to competing amino acids ratio in depressive disorder: relation to efficacy of antidepressive treatments. Acta Psychiatr Scand 72 (suppl 325):1–31, 1985

Møller SE: Tryptophan and tyrosine ratios to neutral amino acids in depressed patients in regard to K_m: relation to efficacy of antidepressant treatments, in Amino Acid Availability and Brain Function in Health and Disease. Edited by Huether G. Berlin, Springer-Verlag, 1988, pp 355–361

Møller SE, Fogh M: Clinical response to ECT: relation to plasma ratios of tryptophan and tyrosine to other large neutral amino acids. Pharmacopsychiatry 21:63, 1988

Møller SE, Kirk L: Nutrients, neurotransmitters, and behavior. Commentary. Integr Psychiatry 5:249–254, 1987

Møller SE, Kirk L, Honoré P: Free and total plasma tryptophan in endogenous depression. J Affective Disord 1:69–76, 1979

Møller SE, Reisby N, Ortmann J, et al: Relevance of tryptophan and tyrosine availability in endogenous and "non-endogenous" depressives treated with imipramine or clomipramine. J Affective Disord 3:231–244, 1981

Møller SE, Honoré P, Larsen OB: Tryptophan and tyrosine ratios to neutral amino acids in endogenous depression: relation to antidepressant response to amitriptyline and lithium + L-tryptophan. J Affective Disord 5:67–69, 1983a

Møller SE, Kirk L, Brandrup E, et al: Tryptophan availability in endogenous depression—relation to efficacy of L-tryptophan treatment. Adv Biol Psychiatry 10:30–46, 1983b

Møller SE, Ødum K, Kirk L, et al: Plasma tyrosine/neutral amino acid ratio correlated with clinical response to nortriptyline in endogenously depressed patients. J Affective Disord 9:223–229, 1985a

Møller SE, Reisby N, Elley J, et al: Biochemical and diagnostic classification and serum drug levels: relation to antidepressive effect of imipramine. Neuropsychobiology 13:160–166, 1985b

Møller SE, de Beurs P, Timmerman L, et al: Plasma tryptophan and tyrosine ratios to competing amino acids in relation to antidepressant response to citalopram and maprotiline. A preliminary study. Psychopharmacology (Berlin) 88:96–100, 1986

Møller SE, Bech P, Bjerrum H, et al: Plasma tryptophan/neutral amino acids ratio in relation to clinical response to paroxetine and clomipramine in patients with major depression. J Affective Disord 18:59–66, 1990

Myers ED: Predicting the response of depressed patients to biological treatment: the dexamethasone suppression test versus clinical judgement. Br J Psychiatry 152:657–659, 1988

Pardridge WM: Kinetics of competitive inhibition of neutral amino acid transport across the blood-brain barrier. J Neurochem 28:103–108, 1977

Prien RF, Blaine JD: Summary and conclusions. Psychopharmacol Bull 20:295–302, 1984

Puzynski S, Rode A, Bidzinski A, et al: Failure to correlate urinary MHPG with clinical response to amitriptyline. Acta Psychiatr Scand 69:117–120, 1984

Reisby N, Gram LF, Bech P, et al: Imipramine: clinical effects and pharmacokinetic variability. Psychopharmacology (Berlin) 54:263–272, 1977

Rihmer Z, Arató M, György S, et al: Dexamethasone suppression test as an aid for selection of specific antidepressant drugs in patients with endogenous depression. Pharmacopsychiatry 18:306–308, 1985

Sauer H, Kick H, Minne HW, et al: Prediction of the amitriptyline response: psychopathology versus neuroendocrinology. Int Clin Psychopharmacol 1:284–295, 1986

Schatzberg AF, Orsulak PJ, Rosenbaum AH, et al: Toward a biochemical classification of depressive disorders. IV. Pretreatment urinary MHPG levels as predictors of antidepressant response to imipramine. Comm Psychopharmacol 4:441–445, 1980

Schildkraut JJ: The catecholamine hypothesis of affective disorders: a review of supporting evidence. Am J Psychiatry 122:509–522, 1965

Schweitzer I, Maguire KP, Gee AH, et al: Prediction of outcome in depressed patients by weekly monitoring with the dexamethasone suppression test. Br J Psychiatry 151:780–784, 1987

Sedman G: Trial of a sustained release form of amitriptyline (Lentizol) in the treatment of depressive illness. Br J Psychiatry 123:69–71, 1973

Sharma IJ, Venkitasubramanian TA, Agnihotri BR: 3-MHPG as a nonpredictor of antidepressant response to imipramine and electroconvulsive therapy. Acta Psychiatr Scand 74:252–254, 1986

Siegel S: Nonparametric Statistics for the Behavioral Sciences. Tokyo, McGraw-Hill Kogakusha, Ltd., 1956

Spiker DG, Edwards D, Hanin I, et al: Urinary MHPG and clinical response to amitriptyline in depressed patients. Am J Psychiatry 137:1183–1187, 1980

Timmerman L, de Beurs P, Tan BK, et al: A double-blind comparative clinical trial of citalopram vs maprotiline in hospitalized depressed patients. Int Clin Psychopharmacol 2:239–253, 1987

van Praag HM: New evidence of serotonin-deficient depressions. Neuropsychobiology 3:56–63, 1977

Wechsler H, Grosser GH, Greenblatt M: Research evaluating antidepressant medications on hospitalized mental patients: a survey of published reports during a five-year period. J Nerv Ment Dis 141:231–239, 1965

Yuwiler A, Oldendorf WH, Geller E, et al: Effect of albumin binding and amino acid competition on tryptophan uptake into brain. J Neurochem 28:1015–1023, 1977

Chapter 6

Effects of Phenylalanine on the Synthesis, Release, and Function of Catecholaminergic Systems

Timothy J. Maher, Ph.D.

Chapter 6

Effects of Phenylalanine on the Synthesis, Release, and Function of Catecholaminergic Systems

The synthesis of the catecholamines dopamine (DA), norepinephrine (NE), and epinephrine (E) depends on the initial conversion of tyrosine to 3,4-dihydroxyphenylalanine (L-dopa). The enzyme that catalyzes this conversion, tyrosine hydroxylase (TOH), requires iron, molecular oxygen, and a tetrahydrobiopterin cofactor. Tyrosine hydroxylase is believed to be the rate-limiting step in catecholamine synthesis since

1. TOH activity in vitro is much lower than that of any of the other enzymes involved in catecholamine biosynthesis (Nagatsu et al. 1964).
2. The apparent K_m for the overall reaction (tyrosine to NE) approaches that for the TOH step (tyrosine to L-dopa) (Levitt et al. 1965).
3. A maximal rate of NE synthesis is obtained with concentrations of tyrosine below 100 μM, whereas significantly higher concentrations of L-dopa or DA are required (Levitt et al. 1965).
4. Inhibition of TOH has dramatic effects on tissue catecholamine levels, whereas inhibition of the other enzymes beyond TOH has only minimal effects (Spector et al. 1967).

Treatments that increase blood and tissue tyrosine levels, e.g., administering pure tyrosine or supplementing it in the diet, normally

These studies were supported in part by grants from the Center for Brain Sciences and Metabolism Charitable Trust and the National Institutes of Neurological and Communicative Diseases and Stroke (NS-21231).

133

fail to alter the basal levels of DA and NE detectably in neurons at rest since the control of their synthesis is believed to be tightly regulated by TOH. This phenomenon suggests that normal, physiologically occurring levels of tyrosine are nearly adequate to saturate TOH and/or that end-product inhibition of TOH (i.e., by its products DA and NE) occurs. The estimated K_m (25–50 μM) of TOH for tyrosine obtained from both in vitro and in vivo studies indicates that TOH is largely (70 to 80%) saturated under resting conditions (Carlsson and Lindqvist 1978). Thus it has been estimated that, under resting conditions, the addition of excess tyrosine could only increase NE synthesis by approximately 14% (Sved 1983). However, when certain catecholamine-containing neurons are activated (i.e., made to fire frequently for sustained periods), the synthesis of catecholamines may become dependent on the availability of tyrosine. During periods of enhanced neuronal activity TOH is believed to be subject to phosphorylation, which causes a conformational change in the enzyme that renders it no longer dependent on the availability of its cofactor, tetrahydrobiopterin, but rather on the availability of its substrate, tyrosine (Lovenberg et al. 1978). Furthermore, when TOH is phosphorylated, a greatly decreased sensitivity to end-product feedback inhibition by NE is observed.

Using procedures that cause discrete groups of catecholamine-containing neurons to fire frequently—increased (Sved et al. 1979) or decreased (Conlay et al. 1981) blood pressure, electrical stimulation of the rat's tail (Reinstein et al. 1984), activation of the retina with light (Fernstrom et al. 1984), or destruction of 75% of the nigrostriatal neurons so that the remaining neurons accelerate their rate of firing in an attempt to compensate for the decreased neurotransmitter release (Melamed et al. 1980)—enhanced tyrosine availability has been demonstrated to increase catecholamine synthesis or release and/or to augment catecholamine-dependent behaviors. Thus, while tyrosine administration increases blood pressure in rats made hypotensive (Conlay et al. 1981) and decreases blood pressure in spontaneously hypertensive rats (Sved et al. 1979), tyrosine administration fails to alter blood pressure significantly in normotensive rats, in which homeostatic mechanisms that utilize catecholamines are not likely to be intensely activated.

Supporting the link between neuronal firing frequency and sensitivity to tyrosine availability are dopaminergic neurons within the prefrontal and cingulate cortices, which appear to lack synthesis-modulating autoreceptors and typically fire at rates well above normal. While these groups of neurons have been shown to be responsive to supplemental tyrosine, neighboring dopaminergic neurons within the olfactory tubercle and corpus striatum, which have a much slower rate of firing than those described above, fail to respond to supplemental tyrosine (Tam et al. 1987). Studies attempting to demonstrate an effect of tyrosine on catecholamine synthesis and/or function are destined to yield negative results if the particular group of neurons being examined are not normally firing, or made to fire, at accelerated rates (Maher 1988; Saraw et al. 1987).

PHENYLALANINE

Tyrosine is traditionally classified as a nonessential amino acid, since it can be formed from phenylalanine by the action of hepatic phenylalanine hydroxylase. Therefore phenylalanine could support catecholamine synthesis by supplying the organism with additional tyrosine. Besides this potential role as a catecholamine precursor, phenylalanine could also interfere with catecholamine synthesis by inhibiting TOH directly (especially in species like humans, which hydroxylate phenylalanine very slowly). Early studies have demonstrated the ability of low concentrations of phenylalanine (25 μM) to inhibit TOH activity by 64% in homogenates of guinea pig heart (Ikeda et al. 1967), while 100 μM phenylalanine inhibited TOH activity by 78% in rat brain homogenates (Nagatsu et al. 1964). Phenylalanine is also well known for its ability to retard the entry of tyrosine into the brain by competing with it for access to the transport carrier for the large neutral amino acids (LNAAs) in the blood-brain barrier (Pardridge 1977). Thus phenylalanine might be expected to have dual dose-dependent effects on catecholamine synthesis; i.e., *supporting* synthesis at low doses (because of its conversion to tyrosine via phenylalanine hydroxylase) while *inhibiting* it at slightly higher concentrations.

The effects of phenylalanine and tyrosine on the release of DA in vitro have been investigated in superfused rat brain striatal slices subjected to electrical stimulation (360 pulses; 12 Hz; 2 milliseconds) (Milner et al. 1986; Milner and Wurtman 1984). Low concentrations of tyrosine (20 to 50 μM) sustained the electrically evoked release of DA in a dose-dependent manner, as reflected by the maintenance of the S2/S1 ratio (the ratio of the release of DA following the second stimulus relative to that following the first). Tissues superfused without tyrosine and similarly stimulated experienced a 14% and 25% reduction in tyrosine and DA levels, respectively. Superfusion with a tyrosine-containing solution (50 μM) completely prevented these reductions. While low concentrations of phenylalanine (50 μM) only partially sustained DA release in the absence of tyrosine, slightly higher concentrations of phenylalanine (200 μM, a value approximately three times that found in plasma of fasting rat) significantly *decreased* the S2/S1 ratio, even when the superfusion was performed in the presence of tyrosine. Phenylalanine's suppression of DA release was not mimicked by another LNAA, leucine.

The ability of phenylalanine to modify catecholamine synthesis and/or release has also been recently investigated using intracerebral microdialysis, a technique that allows for the monitoring of extracellular levels of DA and its major metabolites dihycroxyphenylacetic acid (DOPAC) and homovanillic acid (HVA) in the striatum of rats (During et al. 1988). In these studies a low dose of phenylalanine (200 mg/kg), which in rats elevates the plasma tyrosine ratio more than the plasma phenylalanine ratio, increased DA release by 59%. (The plasma ratio, i.e., the ratio of a particular LNAA divided by the sum of the other competing LNAA, is a useful peripheral index of the transport of the LNAA from blood to brain. The plasma tyrosine ratio can be defined as the plasma tyrosine concentration divided by the summed concentrations of leucine, isoleucine, valine, phenylalanine, and tryptophan.) When phenylalanine was administered in an intermediate dose (500 mg/kg), which elevates both the plasma tyrosine and phenylalanine ratios similarly, DA release was unaltered. However, when a higher phenylalanine dose (1,000 mg/kg) was administered, which increased the plasma phenylalanine ratio more than that of tyrosine, DA release was *decreased* by 26%. No corresponding changes were observed in striatal DOPAC or HVA concentrations with any of the treatments, suggesting that the altered DA synthesis may have occurred largely in a pool that was preferentially released into synapses. Thus, data exist that require consideration of the dual dose-dependent effect of phenylalanine on catecholamine synthesis and/or release when designing studies involving this amino acid.

ASPARTAME

The artificial sweetener aspartame (L-aspartyl-L-phenylalanine methyl ester) serves as an additional source of phenylalanine in the human food chain. Aspartame contains about 56% phenylalanine, and its ingestion rapidly raises blood levels of phenylalanine as well as its other constituents, methanol and aspartic acid. Although the absolute amount of phenylalanine derived from the aspartame in a 12-oz can of diet soft drink is similar to the amount found in various commonly consumed protein sources (e.g., eggs), the changes that occur in the plasma phenylalanine ratio, and thus the probable concentration of phenylalanine in the brain, following aspartame ingestion are actually *opposite* to those observed following protein consumption. This finding is contrary to some published misconceptions that consider only the absolute amount of phenylalanine in a food or additive (Freedman

1987; Yellowlees 1983). This crucial difference results from the presence of other LNAAs in proteins, and their absence from aspartame. An egg, and every other complete protein, contains most if not all of the other 20 naturally occurring amino acids, some of which, like the other LNAAs, compete with phenylalanine and each other for passage across the blood-brain barrier. While absolute plasma phenylalanine levels also increase following aspartame consumption, the plasma phenylalanine ratio increases because aspartame lacks any of the other competing LNAAs (Cabellero et al. 1986). Hence, aspartame is the only readily available dietary constituent that would be expected to increase selectively the entry of phenylalanine into the brain. Additionally, since carbohydrate ingestion is capable of eliciting the release of insulin, which selectively lowers plasma levels of the branched-chain amino acids without having much effect on plasma phenylalanine, if an aspartame-containing beverage is consumed along with a carbohydrate-rich, protein-poor food, its effects on brain phenylalanine will most likely be doubled (Wurtman 1983). Should the hypothetical individual consuming the aspartame with carbohydrate also happen, unknowingly, to be heterozygous for the phenylketonuria (PKU) gene, which causes the conversion of phenylalanine to tyrosine to be significantly reduced, an additional doubling of aspartame's brain effects might be expected to occur (Stegink et al. 1981). Impaired conversion of phenylalanine to tyrosine is also known to occur in patients with iron-deficiency anemia (Lehmann and Heinrich 1986).

POTENTIAL NEUROLOGICAL EFFECTS

Reports of adverse central nervous system effects (including seizures, headache, and mood changes) associated with the consumption of aspartame have increased with the incorporation of this artificial sweetener into foods (Wurtman 1985). It is believed that catecholamines play a role in the suppression of seizure activity in experimental animal models. Animals given drugs that deplete their brains of particular monoamine neurotransmitters or that block the receptor-mediated effects of these transmitters (e.g., reserpine and Ro 4-1284) respectively exhibit greater sensitivity to seizures in many animal models (Jobe et al. 1986). In contrast, treatments thought to act by enhancing monoaminergic neurotransmission (e.g., L-dopa plus a monoamine oxidase inhibitor, or 3,4-dihydroxyphenylserine) apparently protect rodents against the development of various types of experimentally induced seizures. *Low* doses of phenylalanine or aspartame—which, in rodents, raise plasma tyrosine levels more than phenylalanine levels—might be expected to have little or no effect on seizure susceptibility, or even, conceivably, to protect animals against the seizure-promoting effects of drugs like pentylenetetrazol (PTZ) by supplying additional tyrosine and thereby enhancing catecholaminergic neurotransmission. In contrast comparable doses given to humans (who hydroxylate phenylalanine only very slowly), or sufficiently high aspartame doses (which transiently surpass the liver's capacity to hydroxylate the phenylalanine) given to rodents, might enhance seizure susceptibility.

A proconvulsant effect of phenylalanine has been demonstrated in rats exposed to high levels of this amino acid in the diet (Gallagher 1969). Similarly, aspartame has been shown to have a proconvulsant effect in mice treated with PTZ and fluorothyl (Pinto and Maher 1988). At 1,000 and 2,000 mg/kg aspartame, 78% and 100%, respectively, of the animals experienced PTZ-induced seizures, compared with 50% in the water-pretreated control group. Other mice pretreated with a fixed dose (1,000 mg/kg) of aspartame or with water and given various doses (50 to 75 mg/kg) of PTZ 1 hour later exhibited a significant leftward shift of the PTZ dose-response curve (CD$_{50}$ [95% confidence interval] lowered from a control value of 66 [64 to 68] to 59 [56 to 63] mg/kg). Enhanced susceptibility to PTZ-induced seizures was also observed among mice pretreated with equimolar phenylalanine doses but not among animals pretreated with aspartame's other metabolites, aspartic acid or methanol. Coadministration of the LNAA valine with aspartame protected mice from the seizure-promoting effects of the sweetener; in contrast, alanine, an amino acid that does not compete with phenylalanine or the other LNAA for transport into the brain, failed to attenuate aspartame's effect among mice pretreated with equimolar phenylalanine doses but not among animals pretreated with aspartame's other metabolites, aspartic acid or methanol. Some investigators have recently reported similar findings using different convulsants or species (Chiu and Woodbury 1988; Garratini et al. 1988; Kim and Kim 1986), while other investigators fail to see proconvulsant effects of aspartame or phenylalanine using other experimental models (Chiu and Woodbury 1988; Dailey et al. 1987; Garattini et al. 1988; Meldrum and Nanji 1988; Nevins et al. 1986).

FUTURE STUDIES

Experimental designs using techniques that are capable of detecting rather subtle effects on human brain function will be required to assess fully the effects of phenylalanine since the interpretation of studies using intact rodent preparations is confounded by the major phenylalanine metabolizing differences between humans and rodents.

One preliminary study (Spiers et al. 1988) has demonstrated a deleterious effect of chronic (12 days) acceptable daily intake doses (50 mg/kg) of aspartame on tests of cognitive function (think fast and Stroop word reading), while another study using an acute lower dose (15 mg/kg) failed to demonstrate any effects of this artificial sweetener on neuropsychiatric parameters (Lapierre et al. 1988).

Evidence now exists that clearly demonstrates the ability of phenylalanine to affect catecholaminergic processes in several experimental paradigms. When phenylalanine is provided at low concentrations in the absence of tyrosine, or when phenylalanine or aspartame are administered to rodents in doses that elevate brain tyrosine relative to phenylalanine, enhanced catecholaminergic indices have been observed. However, when phenylalanine is present in higher concentrations (two to three times normal), or when phenylalanine or aspartame are administered in doses that elevate brain phenylalanine more than tyrosine, some catecholaminergic indices have been reported to be impaired. With the increasing availability to consumers of aspartame-sweetened foods, the effects of phenylalanine on catecholamine synthesis and related behaviors require further investigation.

REFERENCES

Caballero B, Mahon B, Rohr F, et al: Plasma amino acid levels after single dose aspartame consumption in phenylketonuria. J Pediatr 109:668–671, 1986

Carlsson A, Lindqvist M: Dependence of 5-HT and catecholamine synthesis on precursor amino acid levels in rat brain. Naunyn Schmiedebergs Arch Pharmacol 303:157–164, 1978

Chiu P, Woodbury DM: Effects of aspartame (ASM) and its metabolites on seizure susceptibility in mice. Pharmacologist 30:A119, 1988

Conlay LA, Maher TJ, Wurtman RJ: Tyrosine increases blood pressure in hypotensive rats. Science 212:559–560, 1981

Dailey J, Lasley SM, Frasca J, et al: Aspartame (ASM) is not pro-convulsant in the genetically epilepsy prone rat (GEPR). Pharmacologist 29:142, 1987

During MJ, Acworth IN, Wurtman RJ: An in vivo study of dopamine release in striatum: the effects of phenylalanine, in Dietary Phenylalanine and Brain Function. Edited by Wurtman RJ, Ritter-Walker E. Boston, Birkhauser, 1988, pp 395–403

Fernstrom MH, Volk EA, Fernstrom JD: In vivo tyrosine hydroxylation in the diabetic rat retina: effect of tyrosine administration. Brain Res 298:167–170, 1984

Freedman M: Consumption of aspartame by heterozygotes for phenylketonuria. J Pediatr 110:662–663, 1987

Gallagher BB: Amino acids and cerebral excitability. J Neurochem 16:701–706, 1969

Garratini S, Caccia S, Romano M, et al: Studies on the susceptibility to convulsions in animals receiving abuse doses of aspartame, in Dietary Phenylalanine and Brain Function. Edited by Wurtman RJ, Ritter-Walker E. Boston, Birkhauser, 1988, pp 131–143

Ikeda M, Levitt M, Udenfriend S: Phenylalanine as a substrate and inhibitor of tyrosine hydroxylase. Arch Biochem Biophys 120:420–427, 1967

Jobe PC, Dailey JW, Reigel CE: Noradrenergic and serotoninergic determinants of seizure susceptibility and severity in genetically epilepsy-prone rats. Life Sci 39:775–782, 1986

Kim KC, Kim SH: Studies on the effects of aspartame and lidocaine interaction in central nervous system of mice. Federation Proceedings 46:705, 1986

Lapierre KA, Greenblatt DJ, Goddard JE, et al: Neuropsychiatric effects of aspartame in normal volunteers. J Clin Pharmacol 28:912, 1988

Lehmann WD, Heinrich HC: Impaired phenylalanine-tyrosine conversion in patients with iron-deficiency anemia studied by a L-(^{2}H-5)-phenylalanine-loading test. Am J Clin Nutr 44:468–474, 1986

Levitt M, Spector S, Sjoerdsma A: Elucidation of the rate-limiting step in norepinephrine biosynthesis in the perfused guinea-pig heart. J Pharmacol Exp Ther 148:1–9, 1965

Lovenberg W, Bruckswick EA, Hanbauer I: ATP, cyclic AMP, and magnesium increase the affinity of rat striatal tyrosine hydroxylase for its cofactor. Proc Natl Acad Sci USA 72:2955–2958, 1978

Maher TJ: Tyrosine and brain function. ISI Atlas of Science: Biochemistry 1:150–154, 1988

Melamed E, Hefti F, Wurtman RJ: Tyrosine administration increases striatal dopamine release in rats with partial nigrostriatal lesions. Proc Natl Acad Sci USA 77:4305–4309, 1980

Meldrum B, Nanji N: Lack of effect of large doses of aspartame on photically-induced seizures in the baboon (Papio papio). FASEB J 2:A434, 1988

Milner JD, Wurtman RJ: Release of endogenous dopamine from electrically stimulated slices of rat striatum. Brain Res 301:139–142, 1984

Milner JD, Irie K, Wurtman RJ: Effects of phenylalanine on the release of endogenous dopamine from rat striatal slices. J Neurochem 47:1444–1447, 1986

Nagatsu T, Levitt M, Udenfriend S: Tyrosine hydroxylase: the initial step in norepinephrine synthesis. J Biol Chem 239:2910–2917, 1964

Nevins ME, Arnolde SM, Haigler HJ: Aspartame: lack of effect on convulsant thresholds in mice. Federation Proceedings 45:1096, 1986

Pardridge WM: Regulation of amino acid availability to the brain, in Nutrition and the Brain, Vol 1. Edited by Wurtman RJ, Wurtman JJ. New York, Raven Press, 1977, pp 141–204

Pinto JMB, Maher TJ: Administration of aspartame potentiates pentylenetetrazole- and fluorothyl-induced seizures in mice. Neuropharmacology 27:51–56, 1988

Reinstein DK, Lehnert H, Scott NA, et al: Tyrosine prevents behavioral and neurochemical correlates of an acute stress in rats. Life Sci 34:2225–2231, 1984

Saraw G, Behrens WA, Peace RW, et al: Lack of relationship between dietary tyrosine and sympathetic nervous system activity in rats fed normal protein diets. Nutr Rep Int 35:471–478, 1987

Spector S, Gordon R, Sjoerdsma A: End product inhibition of tyrosine hydroxylase as a possible mechanism for regulation of norepinephrine synthesis. Mol Pharmacol 3:549–555, 1967

Spiers P, Schommer D, Sabounjian L: Aspartame and human behavioral observations, in Dietary Phenylalanine and Brain Function. Edited by Wurtman RJ, Ritter-Walker E. Boston, Birkhauser, 1988, pp 169–178

Steqink LD, Koch R, Blaskovics ME, et al: Plasma phenylalanine levels in phenylketonuric heterozygous and normal adults administered aspartame at 34 mg/kg body weight. Toxicology 20:81–90, 1981

Sved AF: Precursor control of the function of monoaminergic neurons, in Nutrition and the Brain. Edited by Wurtman RJ, Wurtman JJ. New York, Raven Press, 1983, pp 223–275

Sved AF, Fernstrom JD, Wurtman RJ: Tyrosine administration reduces blood pressure and enhances brain norepinephrine release in spontaneously hypertensive rats. Proc Natl Acad Sci USA 76:3511–3514, 1979

Tam SY, Ono N, Roth RH: Precursor control and influence of aspartame on midbrain dopamine neurons, in Amino Acids in Health and Disease:

New Perspectives, Vol 55. Edited by Kaufman S. New York, Alan R Liss, 1987, pp 421–435

Wurtman RJ: Neurochemical changes following high-dose aspartame with dietary carbohydrates. N Engl J Med 309:429–430, 1983

Wurtman RJ: Aspartame: possible effect on seizure susceptibility. Lancet 2:1060, 1985

Yellowlees H: Aspartame. Br Med J 287:162–163, 1983

Chapter 7

Plasma Phenylalanine: A Measure of Tardive Dyskinesia Vulnerability in Schizophrenic Males

Mary Ann Richardson, Ph.D.
Raymond Suckow, Ph.D.
Rowland Whittaker, R.N.
Howard Kushner, Ph.D.
William A. Boggiano, M.D.
Istvan Sziraki, Ph.D.

Chapter 7

Plasma Phenylalanine: A Measure of Tardive Dyskinesia Vulnerability in Schizophrenic Males

Tardive dyskinesia (TD) is an abnormal movement disorder that is considered to be secondary to neuroleptic treatment. The syndrome is seen in large numbers of chronic psychiatric patients, mainly schizophrenic patients, for whom neuroleptic drugs are the prime treatment medium.

Epidemiological data evidence an etiological role for neuroleptics in the disorder. This role is demonstrated by the contrast of data from studies defining the prevalence of TD in neuroleptic-treated versus nonneuroleptic-treated populations. In this work, the prevalence of TD is shown to be $3\frac{1}{4}$ times greater in the neuroleptic-treated populations (Jeste and Wyatt 1981). Despite this demonstrated etiological role for neuroleptics, there is a marked lack of consistent and robust findings in the study of drug history variables as risk factors for TD development (e.g., quantity and type of neuroleptic treatment, antiparkinson use, antidepressant use) other than the variable of the length of time since first neuroleptic treatment (Branchey et al. 1983; Simpson et al. 1978; Wojcik et al. 1980). This lack of potency for drug history variables has turned research attention to patient vulnerability factors, which have often been linked to proposed pathophysiological mechanisms for the disorder. Some hypotheses implicate dopaminergic (Bowers and Glazer 1987), noradrenergic (Jeste et al. 1984; Kaufman et al. 1986; Wagner et al. 1982), or

This work was supported in part by a National Institute of Mental Health grant (R03 MH-40629-01A). The authors wish to acknowledge the assistance of Frances Simpson, Brenda Katof, Audrey Meister, Vipin Trevidi, Margaret Bevans, and Jill Joyce.

GABA-ergic (Fibiger and Lloyd 1984; Stahl et al. 1985; Thaker 1987) mechanisms in the development of TD. Biological risk factors in patient populations have been sought to verify these hypotheses. Plasma, urine, and cerebrospinal fluid (CSF) measures of dopaminergic, noradrenergic, and serotoninergic metabolism have been examined as correlates of TD. The paucity of consistent findings in these investigations has been disappointing. These collected data across studies do not demonstrate an underlying dopaminergic dysfunction (the most commonly accepted theory), provide only inconsistent support for a noradrenergic pathophysiologic construct, and present no data for serotonin involvement (Jeste et al. 1981, 1984; Kaufman et al. 1986; Markianos et al. 1983; Moore et al. 1983; Tripodianakis et al. 1983; Wagner et al. 1982).

Recently, a new direction in TD etiology/pathophysiology research, a hypothesized vulnerability model for TD based on the metabolism of the large neutral amino acid (LNAA) phenylalanine, has been developed by the first author. This hypothesis was generated from the results of two studies in which the association of patient characteristics with TD in two markedly different neuroleptic-treated populations (mentally retarded and young male schizophrenics) was investigated.

The first study (Richardson et al. 1986) was a point prevalence investigation of TD conducted among residents of a state facility for the mentally retarded. The study population consisted of 211 patients (139 males and 72 females) with a mean age of 35.2 years and a mean length of time since first admission of 27.3 years The mean IQ for the group was 15.5. Sixty percent of the patients were being treated with neuroleptics at the time of evaluation. The prevalence of TD using the study criterion measure was 29.9%. This criterion measure was a dichotomy based on a global score derived from a subset of items from the Simpson Abbreviated Dyskinesia Scale (ADS) (Simpson et al. 1979). These items were lip chewing and lateral jaw movements, bon bon sign, tongue protrusion, choreoathetoid movements of the tongue, cheek puffing, and choreoathetoid movements of the fingers/wrists and ankles/toes. A stepwise discriminant function analysis demonstrated that the following variables were significant discriminators for TD (accuracy rate = 77.3%, jackknifed rate = 76.3%): age (positive), DSM-II (American Psychiatric Association 1968) category of Retardation Due to Disorders of Metabolism, Growth or Nutrition (positive); DSM-II category of Retardation Due to Psychosocial Deprivation (negative); sex (female). The following independent variables were not significant discriminators for TD:

- Length of time in treatment.
- Number of days on neuroleptics in the 5 years prior to evaluation.
- Neuroleptic status at evaluation.
- Neuroleptic dose at evaluation.
- Continuous number of days off neuroleptics prior to evaluation.

The lack of significant positive association between TD and the other diagnostic categories, such as retardation due to trauma, physical agent, or gross brain disease, speaks against a strong association between the disorder and structural organic conditions. The modal condition in the significantly positively associated diagnostic category of Disorders of Metabolism, Nutrition and Growth was phenylketonuria (PKU). The TD prevalence among the PKU patients was 86%, while the TD prevalence in the remainder of the study population was 27%.

PKU is an autosomal recessive disease caused by a defect in the enzyme system that catalyzes the conversion of phenylalanine to tyrosine. Classical PKU is due to a deficiency of phenylalanine-4-hydroxylase; atypical PKU is due to a deficiency of dihydropteridine reductase, which is necessary for the regeneration of the essential cofactor L-erythro-5,6,7,8-tetrahydrobiopterin (BH_4). BH_4 functions as the natural cofactor in the hydroxylation of phenylalanine to tyrosine, tyrosine to dihydroxyphenylalanine (dopa), and tryptophan to serotonin. The symptoms of PKU are mental retardation, seizures, spasticity, and EEG irregularities. The behavioral characteristics of the disorder are anxiety, restlessness, night terrors, destructiveness, noisiness, hyperactivity, irritability, and uncontrollable temper tantrums. Because of the enzyme defect in PKU, the phenylalanine plasma level in untreated PKU patients increases up to 50 mg/dl (Curtius et al. 1981). These high levels of phenylalanine saturate the blood-brain barrier L-carrier transport system, the system for the uptake of the neutral amino acids into the brain. Because of a particular affinity of the blood-brain barrier for phenylalanine in preference to the other LNAAs, elevations in plasma phenylalanine block the uptake of the other LNAAs such as tyrosine and tryptophan into the brain. High levels of phenylalanine have also been shown to inhibit tyrosine hydroxylase (Ikeda et al. 1967). These reportedly high levels of phenylalanine in PKU have also been shown to produce excessive quantities of the derivatives of phenylalanine: phenylethylamine (PEA) (Oates et al. 1963) and phenylpyruvic acid, phenyllactic acid, and phenylacetic acid (PAA) (Knox and Hsia 1957). Fellman (1956) has demonstrated (in vitro) that phenylpyruvic acid, phenyllactic acid, and PAA inhibit dopa decarboxylase. Measurement of monoamines and their metabolites in PKU patients has shown the high phenylalanine levels of PKU to decrease plasma, urine, and CSF levels of 5-hydroxyindoleacetic acid (5-HIAA), homovanillic acid (HVA), 3-methoxy-4-hydroxyphenylglycol, vanilmandelic acid, dopamine, noradrenaline, epinephrine, and 5-hydroxytryptamine (5-HT) (Butler et al. 1981; Cession-Forsion et al. 1966; Curtius et al. 1981; Nadler and Hsia 1961; Pare et al. 1957; Weil-Malherbe 1955). In the Curtius et al. (1981) study, dopamine formation required phenylalanine plasma levels of 25 mg/dl for inhibition; serotonin, however, required levels of only 8 mg/dl for inhibition. Autopsy PKU work demonstrates high phenylalanine associated with a 30 to 40% decrease of tyrosine and tryptophan (McKean 1972). A study conducted on older, treated PKU patients measuring plasma phenylalanine and urinary dopamine found that in 9 out of 10 patients, an inverse relationship existed between plasma phenylalanine and urinary dopamine. In the same study, urinary serotonin fell during phenylalanine loading in six patients (Krause et al. 1985).

The deficits of PKU have a three-pronged etiology: 1) a saturation of brain cells by phenylalanine; 2) a decrease in the brain levels of the monoamines caused by phenylalanine competing with the monoamine precursors (tyrosine and tryptophan) for uptake into the brain; and 3) a decrease in the brain levels of the monoamines caused by the inhibition by phenylalanine of the monoamine-synthesizing enzymes.

The second study that led to the phenylalanine/TD hypothesis (Richardson et al. 1985) was conducted among young male inpatients with schizophrenia (ages 18 to 44) at a New York State psychiatric center. The patients were placed into either a case group ($n = 16$ with TD) or a control group ($n = 16$ without TD) using the criterion measure described above. Cases and controls were age matched so that there were no significant age differences between the two groups. Subjects were assessed for mental status and psychopathology with the Hamilton Depression Rating Scale (HDRS) (Hamilton 1960), the Brief Psychiatric Rating Scale (BPRS) (Overall and Gorham 1962), the Affective Flattening Scale (AFS) (Andreasen 1979), and the Mini-Mental State Exam (MMSE) (Folstein et al. 1975). Symptoms of interest collected during the chart review and diagnostic interview were also converted into independent variables. The total BPRS score and the number of years since first admission (duration variable) were the only two significant positive discriminators for the presence of TD according to the stepwise discriminant function analysis (accuracy rate = 90.6%, jackknifed rate = 81.3%). Other variables that did not prove significant were length of current admission, chlorpromazine (CPZ) equivalent dose at evaluation, and total scores for the HDRS. Scores from the AFS and MMSE had not been included as variables in the multivariable analysis because they were not associated with TD in a preliminary univariate significance screening. To study the BPRS association further, a second discriminant function analysis was performed on the five BPRS factors (anxiety-depression, anergia, thought disturbance, activation, hostility-suspiciousness) and the duration variable. Along with duration, the BPRS factors of activation and hostility-suspiciousness proved to be significant positive discriminators for TD (accuracy rate = 81.3%, jackknifed). Thus, from the study analyses, the individual BPRS items that were significant discriminators for TD were tension, mannerisms and posturing, and excitement and hostility. A χ^2 analysis demonstrated a significant association between a history of DSM-III–defined (American Psychiatric Association 1980) manic symptoms and TD ($\chi^2 = 6.36$; $P = .01$), while DSM-III–defined depressive symptoms were not associated. The "inappropriate affect" item from the AFS was also significantly associated with TD ($u = 54.0$, $P < .01$). This study, then, demonstrated that young male schizophrenic patients who developed signs of TD were also characterized by a symptom constellation with affective qualities in the direction of mania.

Relevant to the phenylalanine/TD hypothesis are reports of abnormally high urinary levels of PEA in patients with manic symptoms (Fischer and Heller 1972). In one study, five women with bipolar affective disorders periodically excreted very high urinary concentrations of PEA (Karoum et al. 1982). The PEA excretion rate did not correlate with mood ratings but seemed to be a trait condition. The high PEA excretors were characterized by inappropriate affect. In another study, three women with rapidly cycling bipolar disorders who also manifested depression-dependent dyskinesias showed abnormally high levels of PEA (Linnoila et al. 1983). PEA is the decarboxylated product of phenylalanine. The administration of phenylalanine produces a threefold increase in the concentration of PEA in the rat brain (Boulton and Juorio 1982). PEA is a sympathomimetic amine that is a chemical congener of amphetamine and is oxidized by monoamine oxidase (type B) (Antelman et al. 1977). It is a lipophilic compound that readily passes through the blood-brain barrier (Oldendorf 1971). The concentration of PEA is generally low; it has a relatively short half-life and a high turnover rate (Boulton and Juorio 1982). The metabolism rate of PEA in the human brain is similar to that of serotonin (Young et al. 1982). Systemic injection of low doses of PEA to rats produced an instantaneous and dose-dependent inhibition of the firing rate of locus coeruleus neurons and dopamine-containing neurons in the substantia nigra (Oreland et al. 1985). It has been proposed that PEA may act as a neuromodulator (Boulton and Juorio 1982).

In view of the following points, a study was undertaken to test a hypothesis that phenylalanine metabolism may be associated with TD:

• The associations of TD with PKU.
• The disturbances of affect in PKU.
• The association of TD with manic symptoms in schizophrenic patients.
• The known disorder of phenylalanine metabolism in PKU.
• The suggestion of abnormalities of the phenylalanine metabolic pathway in patients with mania.

METHODS

Subjects

All male inpatients of a state psychiatric center between December 1984 and March 1986, ages 18 to 44, and carrying a chart diagnosis of schizophrenia were screened for study inclusion if there were no fewer than 3 years nor more than 20 years since their first neuroleptic treatment. Patients were excluded if they were incompetent to give informed consent, refused consent, were non-English-speaking, were diabetic, were mute, were deaf, or failed to satisfy criteria (DSM-III or Research Diagnostic Criteria) for schizophrenia after a lifetime symptom screening and diagnostic interview.

Procedure

All subjects arrived at the institute at 6:30 A.M. having fasted from 10 P.M. the previous evening. Fasting bloods were collected for plasma amino acid and PEA determinations. A breakfast was then served as a protein loading. The breakfast consisted of one 8-ounce glass of milk, 2½ eggs, ⅓ cup of grated cheddar cheese, 6 ounces of ham, one 4-ounce glass of orange juice, and one waffle. The protein, fat, carbohydrate, and amino acid breakdown of the meal was as follows (Pennington and Church 1985):

    80.8 gm—protein
    60.8 gm—fat
    42.1 gm—carbohydrates
    1,614 mg—arginine
    460 mg—cystine
    916 mg—histidine
    4,359 mg—isoleucine
    6,812 mg—leucine
    6,314 mg—lysine

2,065 mg—methionine
3,551 mg—phenylalanine
3,345 mg—threonine
 906 mg—tryptophan
1,476 mg—tyrosine
4,718 mg—valine

Bloods were drawn 2 hours after each subject finished eating for postloading estimates of amino acids and PEA.

Patients were evaluated at the experimental session for neuroleptic side effects and further evaluated at subsequent rating sessions for both neuroleptic side effects and psychiatric status. The ADS was used to evaluate patients for TD. The TD designation was based on the same dichotomous criterion measure of a mild to severe global rating on a subset of ADS items as described above. A severity score was also calculated from the sum of the items in that same subset of the ADS. Psychiatric status was evaluated by use of the BPRS. These data were analyzed in total score form (items 1 to 18). In addition to the side effect and clinical status, among the other variables quantified for study were age at first neuroleptic treatment and length of time since first neuroleptic treatment (duration).

Plasma Amino Acid Analysis

Sample Preparation. Analysis of amino acids in plasma was accomplished by a technique involving high-performance liquid chromatography preceded by precolumn derivation of the samples by phenylisothiocyanate (PITC) (Bidlingmeyer et al. 1984).

Plasma samples (0.25 ml) were deproteinized by mixing with equal volumes of acetonitrile followed by centrifugation. The clear protein-free supernatant was mixed with two volumes of methyl-t-butyl ether and centrifuged, and the ether layer was discarded. The aqueous layer was evaporated to dryness under vacuum. The residue was redissolved in a mixture of ethanol:triethylamine:water (2:1:2) (40 μl) and again redried. Derivation was initiated by adding a (50 μl) freshly prepared reagent consisting of ethanol, triethylamine, water, and PITC in the ratio 7:1:1:1. After 20 minutes at room temperature, the entire reagent mix was removed under vacuum. (The residue may be stored at −20°C for several weeks without any significant degradation of the amino acids.) This residue was redissolved in a phosphate buffer containing 5% acetonitrile prior to chromatographic analysis. Standards of free amino acids in 5% albumin bovine were carried through the entire procedure.

Chromatography. The separation of the amino acids was performed by liquid chromatography. The solvent delivery system consisted of two Model 6000A pumps and a Model 440 fixed-wavelength ultraviolet detector (254 nm). The solvent gradient was controlled with a Model 660 Gradient Programmer, and the samples were injected by a Model 710B WISP auto injector (all equipment by Waters Assoc.). The column temperature was controlled at 39 °C by an aluminum column block connected to a Haake Circulating Water Bath. The column was an application-specified reversed-phase PICO-TAG column, 3.9 × 150 mm.

The solvent system consisted of two eluents: an aqueous buffer and 60% acetonitrile in water. The typical buffer (A) was 0.14 M sodium acetate containing 0.5 ml of triethylamine per liter and titrated to pH 6.35 with glacial acetic acid with 6% acetonitrile. A gradient used for the separation consisted of 3% B transversing to 51% B at a flow rate of 1.1 ml/minute for 13 minutes using convex curve #5 on the Model 660 Gradient Programmer. An additional step at 100% B was used to wash the column prior to returning to the initial conditions for reequilibration.

Peak heights of the amino acids on the chromatogram were compared to the peak heights of the amino acid standards processed simultaneously using norleucine as the internal standard. A standard calibration curve developed from the standards was used to quantify each amino acid.

To account for the competition among the LNAAs for uptake into the brain from the L-carrier transport system, ratios of plasma levels of phenylalanine, tyrosine, and tryptophan to the other LNAAs were calculated in order to estimate the rates at which each amino acid entered the brain (Fernstrom and Wurtman 1972; Pardridge and Choi 1986). For example, the phenylalanine ratio was calculated as follows:

$$\text{Phenylalanine ratio} = \frac{\text{Phenylalanine plasma level}}{\text{Sum of the plasma levels of histidine, threonine, tyrosine, valine, isoleucine, leucine, and tryptophan}}$$

Plasma PEA Analysis. PEA was quantified using a GC/MS negative chemical ionization procedure. Deuterium-labeled PEA was used as an internal standard. The PEA was separated from plasma or tissue by liquid/liquid extraction then derivatized using pentafluorobenzoyl chloride. After removal of excess reagent, the derivatized material was injected onto a 30-meter DB-17 megabore column at 210 °C with methane as carrier and ionization gas. Negative chemical ionization was performed with the source temperature at 150 °C and pressure of 1 torr. Fragmentation was minimal with the derivative with the major ion at amu 295 and 298 for PEA and deuterated PEA, respectively. The method is highly sensitive, with a limit of detection of 20 ng/ml and linear through 2 ng/ml. In 50 drug-free patients not associated with this study, using this procedure, the mean plasma PEA level was 85.3 ng/ml ± 27.8.

Statistical Analysis

Each individual variable (risk factor) was screened for possible significance with respect to TD by successive univariate t tests or the Behrens-Fischer t test, depending on which was appropriate. Also, possible dependencies among the variables were tested by correlational analysis. As is well known, such a procedure may produce "significant" P values by chance alone. The P values from these successive t tests must, therefore, be treated with caution.

Logistic models with TD status as the dependent variable were prepared to test simultaneously variables (risk factors) determined to be individually significantly associated with TD in the t-test analyses. The forced-entry approach was used for the logistic regression analyses. The number of variables used closely corresponded to 1/10 the number of observations. It was believed that the forced-entry procedure and the limitation of variables to four or five would maximize the rigor of the logistic analyses and would enhance the interpretation of the risk factors determined from these models. Two models were used: the first logistic model analyzed variables pertinent to the fasting condition; the second model employed significant variables appropriate for the postloading condition.

RESULTS

Amino Acid Correlations

A correlational analysis performed on the amino acids and PEA data (Table 7-1) shows a significant positive association between loading levels of PEA and phenylalanine. PEA also shows some significant correlations with tyrosine and with tryptophan. Phenylalanine, however, shows a stronger picture than PEA of associations with tyrosine and tryptophan. Correlations between tyrosine and tryptophan were stronger subsequent to the protein loading.

TD–Yes vs. TD–No

Patients were classified by TD status: Yes ($n = 34$) or No ($n = 19$). T tests were used to test differences between the two TD status groups on the continuous variables. The demographic variable, age at first neuroleptic treatment, showed significant differences between the TD-Yes/No groups, with the TD-Yes group showing a higher mean value ($t = 2.04$, df = 51, $P \le .05$). The BPRS total score also showed significant differences between the groups ($t = 3.21$, df = 28.6, $P \le .01$), with the TD-Yes group showing higher mean values (means: TD-Yes = 46.1; TD-No = 36.7).

The postloading phenylalanine level was significantly higher for the TD-Yes group (means: TD-Yes = 68.7; TD-No = 58.6; $t = 2.20$, df = 51, $P \le .05$). No significant difference between groups was seen for the following levels: phenylalanine-fasting, PEA-fasting and postloading, tyrosine-fasting and postloading, and tryptophan-fasting and postloading. Significant phenylalanine/LNAA ratio differences for TD status groups were found both fasting and postloading, with higher ratios for TD-Yes (fasting—means: TD-Yes = .0792; TD-No = .0711; $t = 2.17$, $P \le .05$, postloading—means: TD-Yes = .0782; TD-No = .0658; $t = 3.20$, df = 46, $P \le .01$). No significant differences between TD groups were seen for the tyrosine- or the tryptophan-fasting or postloading ratios.

Table 7-1.   Significant plasma amino acid level correlations in male schizophrenic patients

	PEA-PL	PHE-F	PHE-PL	TYR-F	TYR-PL	TRP-F	TRP-PL
PEA-F							
r	.74		.31			.35	.40
P	[.0001		[.05			[.05	[.01
PEA-PL							
r		.35	.43		.31		.39
P		[.05	[.005		[.05		[.01
PHE-F							
r			.66	.63	.54	.35	.40
P			[.0001	[.0001	[.01	[.01	[.01
PHE-PL							
r				.37	.69		.56
P				[.01	[.0001		[.0001
TYR-F							
r					.59	.49	.47
P					[.0001	[.001	[.001
TYR-PL							
r						.37	.69
P						[.01	[.0001
TRP-F							
r							.65
P							[.0001

Note.   PHE = phenylalanine. TYR = tyrosine. TRP = tryptophan. F = fasting levels. PL = post-protein-loading levels.
Source.   Reprinted from Richardson et al. 1988, with permission from Humana Press. Copyright 1988.

In addition to being associated with TD status, phenylalanine postloading was also weakly but significantly correlated with TD severity ($r = .27$, $P \le .05$), as were phenylalanine ratios fasting and postloading (phenylalanine ratio fasting—$r = .28$, $P \le .05$; phenylalanine ratio postloading—$r = .33$, $P \le .05$). A trend for an association with TD severity was seen for PEA postloading ($r = .23$, $P = .10$).

The risk factors (variables) found to be significant in both logistic regression models, thereby demonstrating significant independent association with TD, were the phenylalanine/LNAA ratio fasting and postloading, age at first neuroleptic treatment, and BPRS total score.

DISCUSSION

This study represents the first attempt to study amino acid metabolism in relation to TD status or TD severity. The significant positive association of phenylalanine postloading plasma levels with TD and the independent ability of the phenylalanine/LNAA ratio to explain TD status, with higher phenylalanine ratios for TD patients in both the fasting and postloading conditions, have implications for the balance of brain amino acids in TD. The significant positive association of phenylalanine with TD severity, although modest, further strengthens the phenylalanine/TD association. The fact that PEA levels did not differentiate TD status, while phenylalanine did, suggests there may be some dissociation between the amine and the amino acid as regards the pathophysiology of TD. The plasma level of PEA, however, did tend to be associated with TD severity. The possibility cannot be discounted, further, that metabolites of PEA such as PAA would have shown an association with TD status.

Phenylalanine-induced alterations in brain amino acid availability may lead to changes in neurotransmitter metabolism (dopaminergic, noradrenergic, serotonergic) (Pardridge and Choi 1986). Looking to our PKU model, because of the competition at the blood-brain barrier among phenylalanine, tyrosine, and tryptophan and because of the demonstrated inhibition by phenylalanine excesses of dopa decarboxylase, tyrosine hydroxylase, and tryptophan hydroxylase, it can be expected that high phenylalanine is associated with decreased plasma, urine, and CSF levels of dopamine, noradrenaline, serotonin, and their metabolites. Of interest here is that markedly lower plasma levels of phenylalanine than those found in PKU have been demonstrated in schizophrenic patients to lead to lower levels of tyrosine, tryptophan, HVA, and 5-HIAA. A study in schizophrenic patients found high plasma phenylalanine to be associated with decreased CSF tyrosine, tryptophan, HVA, and 5-HIAA, which was not true for the normal subjects studied (Bjerkenstedt et al. 1985).

Adding to the above-mentioned report of raised phenylalanine levels in schizophrenic patients and to our finding of higher phenylalanine levels and ratio in a neurological disorder (TD) within a schizophrenic population is some work done in neurological populations. Phenylalanine (CSF) has been studied across the disorders of Parkinson's disease, senile dementia, dystonia musculorum deformans, choreathetosis, essential and hereditary tremor, and motor neuron disease. Phenylalanine was found to be significantly increased across all these conditions, leading us to conclude that victims of these neurological disorders have a common biochemical defect in their amino acid transport systems (de Belleroche et al. 1984; Lakke and Teelken 1976). Low levels of BH4 as seen in atypical PKU have also been implicated in choreoathetosis, Parkinson's disease, and senile dementia (Leeming et al. 1979; Smith et al. 1986; Yamaguchi and Nagatsu 1983).

The hypothesis of a role for phenylalanine in the vulnerability to develop TD and its pathophysiology can be extended to postulate that the antecedent condition of chronically higher levels of phenylalanine for TD-Yes over TD-No may lead to chronically lower levels of dopamine, noradrenaline, and serotonin for that group. Thus, a neurochemical substrate may be created that can, with the precipitant of chronic neuroleptic treatment, lead to the outcome event of TD. Perhaps of more practical clinical importance than etiological definition, these findings can readily lead to the generation of treatment strategies for TD based on plasma amino acid manipulation (as in PKU). The success of treatment can then perhaps lead to the development of prophylactic strategies for the disorder in risk populations.

REFERENCES

American Psychiatric Association: Diagnostic and Statistical Manual of Mental Disorders, 2nd Edition. Washington, DC, American Psychiatric Association, 1968

American Psychiatric Association: Diagnostic and Statistical Manual of Mental Disorders, 3rd Edition. Washington, DC, American Psychiatric Association, 1980

Andreasen NC: Affective flattening and the criteria for schizophrenia. Am J Psychiatry 136:944–947, 1979

Antelman SM, Edwards DJ, Lin M: Phenylethylamine: evidence for a direct, postsynaptic dopamine receptor stimulating action. Brain Res 127:317–322, 1977

Bidlingmeyer BA, Cohen SA, Farvin TL: Rapid analysis of amino acids using pre-column derivatization. J Chromatogr 336:93–104, 1984

Bjerkenstedt L, Edman G, Hagenfeldt L, et al: Plasma amino acids in relation to cerebrospinal fluid monoamine metabolites in schizophrenic patients and healthy controls. Br J Psychiatry 147:276–282, 1985

Boulton AA, Juorio AV: Brain trace amines, in Handbook of Neurochemistry, 2nd Edition, Vol. I: Chemical and Cellular Architecture. Edited by Lajtha A. New York, Plenum Press, 1982, pp 189–222

Bowers MB Jr, Glazer WM: Is there a presynaptic tardive dyskinesia? (letter) J Clin Psychopharmacol 7:57–58, 1987

Branchey M, Richardson MA, Siegel C: Psychotropic drug history and tardive dyskinesia. Psychopharmacol Bull 19:120–122, 1983

Butler IJ, O'Flynn ME, Seifert WE, et al: Neurotransmitter defects and treatment of disorders of hyperphenylalaninemia. J Pediatr 98:729–733, 1981

Cession-Forsion A, Vandermeulen R, Dodinval P, et al: Elimination urinaire de l'adrenaline, de la noradrenaline, et de l'acide VMA chez enfants oligophienes phenylpyruviques. Pathol Biol 14:1157–1159, 1966

Curtius H-Ch, Niederwieser A, Viscontini M, et al: Serotonin and dopamine synthesis in phenylketonuria. Adv Exp Med Biol 133:277–287, 1981

de Belleroche J, Recordati A, Clifford Rose F: Elevated levels of amino acids in the CSF of motor neuron disease patients. Neurochem Pathol 2:1–6, 1984

Fellman JH: Inhibition of DOPA decarboxylase by aromatic acids associated with phenylpyruvic oligophrenia. Proc Soc Exp Biol Med 98:413–414, 1956

Fernstrom JD, Wurtman RJ: Brain serotonin content: physiological regulation by plasma neutral amino acids. Science 178:414–416, 1972

Fibiger HC, Lloyd KG: Neurobiological substrates of tardive dyskinesia: the GABA hypothesis. TINS, Dec 1984, pp 462–464

Fischer E, Heller B: Phenylethylamine as a neurohumoral agent in brain. Behavioral Neuropsychiatry 4:8–11, 1972

Folstein MF, Folstein SE, McHugh PR: Mini-mental state: a practical method for grading the cognitive state of patients for the clinician. J Psychiatr Res 12:189–198, 1975

Hamilton M: A rating scale for depression. J Neurol Neurosurg Psychiatry 23:56–62, 1960

Ikeda M, Levitt M, Udenfriend S: Phenylalanine as substrate and inhibitor of tyrosine hydroxylase. Arch Biochem Biophys 120:420–427, 1967

Jeste DV, Wyatt RJ: Changing epidemiology of tardive dyskinesia: an overview. Am J Psychiatry 138:297–309, 1981

Jeste DV, DeLisi LE, Zalcman S, et al: A biochemical study of tardive dyskinesia in young male patients. Psychiatry Res 4:327–331, 1981

Jeste DV, Doongajl DR, Linnoila M: Elevated cerebrospinal fluid noradrenaline in tardive dyskinesia. Br J Psychiatry 144:177–180, 1984

Karoum F, Linnoila M, Potter WZ, et al: Fluctuating high urinary phenylethylamine excretion rate in some bipolar affective disorder patients. Psychiatry Res 6:215–222, 1982

Kaufman CA, Jeste DV, Shelton RC, et al: Noradrenergic and neuroradiological abnormalities in tardive dyskinesia. Biol Psychiatry 21:799–812, 1986

Knox WE, Hsia DY: Pathogenetic problems in phenylketonuria. Am J Med 22:687, 1957

Krause W, Halminski M, McDonald L, et al: Biochemical and neuropsychological effects of elevated plasma phenylalanine in patients with treated phenylketonuria. J Clin Invest 75:40–48, 1985

Lakke JPWF, Teelken AW: Amino acid abnormalities in cerebrospinal fluid of patients with parkinsonism and extrapyramidal disorders. Neurology 26:489–493, 1976

Leeming RJ, Blair JA, Melikian V: Biopterin derivatives in senile dementia (letter). Lancet, Jan 27, 1979, p 215

Linnoila M, Karoum F, Cutler NP, et al: Temporal association between depression-dependent dyskinesias and high urinary phenylethylamine output. Biol Psychiatry 18:513–517, 1983

Markianos M, Tripodianakis J, Garelis E, et al: Neurochemical studies on tardive dyskinesia. II. Urinary methoxyhydroxyphenylglycol and plasma dopamine-β-hydroxylase. Biol Psychiatry 18:347–354, 1983

McKean CM: The effects of high phenylalanine concentrations on serotonin and catecholamine metabolism in the human brain. Brain Res 47:469–476, 1972

Moore DC, Glazer WM, Bowers MB, et al: Tardive dyskinesia and homovanillic acid. Biol Psychiatry 18:1393–1402, 1983

Nadler HL, Hsia DY: Epinephrine metabolism in phenylketonuria. Proc Soc Exp Biol Med 107:721–723, 1961

Oates JA, Nirenberg PZ, Jepson JB, et al: Conversion of phenylalanine to phenylethylamine in patients with PKU. Proc Soc Exp Biol Med 112:1078–1081, 1963

Oldendorf WH: Brain uptake of radiolabelled amino acids, amines and hexoses after arterial infusion. Am J Physiol 221:1629–1639, 1971

Oreland L, Lundberg P-A, Engberg G: Central effects of tyramine and PEA, in Neuropsychopharmacology of the Trace Amines: Experimental and Clinical Aspects. Edited by Boulton AA, Bieck PR, Maitre L, et al. Clifton, NJ, Humana Press, 1985, pp 201–213

Overall JE, Gorham DR: The brief psychiatric rating scale. Psychol Rep 10:799–812, 1962

Pardridge WM, Choi TB: Neutral amino acid transport at the human blood-brain barrier. Federation Proceedings 45:2073–2078, 1986

Pare CMB, Sandler M, Stacey RS, et al: 5-Hydroxytryptamine deficiency in phenylketonuria. Lancet 1:551–553, 1957

Pennington JAT, Church HN (eds): Food Values of Portions Commonly Used, 14th Edition. New York, Harper & Row, 1985

Richardson MA, Pass R, Bregman Z, et al: Tardive dyskinesia and depressive symptoms in schizophrenics. Psychopharmacol Bull 21:130–135, 1985

Richardson MA, Haugland G, Pass R, et al: The prevalence of tardive dyskinesia in a mentally retarded population. Psychopharmacol Bull 22:243–249, 1986

Richardson MA, Suckow R, Whittaker R, et al: Phenylalanine, phenylethylamine and tardive dyskinesia in psychiatric patients, in The Trace Amines: Their Comparative Neurobiology and Clinical Significance. Edited by Boulton AA, Juorio AV, Downer RGH. Clifton, NJ, Humana Press, 1988

Simpson GM, Lee JH, Zoubok B, et al: A rating scale for tardive dyskinesia. Psychopharmacology (Berlin) 64:171–179, 1979

Simpson GM, Varga E, Lee JH, et al: Tardive dyskinesia and psychotropic drug history. Psychopharmacology 58:117–124, 1978

Smith I, Leeming RJ, Cavanagh NP, et al: Neurological aspects of biopterin metabolism. Arch Dis Child 61:130–137, 1986

Stahl SM, Thornton JE, Simpson ML, et al: Gamma-vinyl-GABA treatment of tardive dyskinesia and other movement disorders. Biol Psychiatry 20:888–893, 1985

Thaker GK, Tamminga CA, Alphs LD, et al: Brain γ-aminobutyric acid abnormality in tardive dyskinesia. Arch Gen Psychiatry 44:522–529, 1987

Tripodianakis J, Markianos M, Garelis E, et al: Neurochemical studies of tardive dyskinesia. I. Urinary homovanillic acid and plasma prolactin. Biol Psychiatry 18:337–345, 1983

Wagner RL, Jeste DV, Phelps BH, et al: Enzyme studies in tardive dyskinesia. I. One-year biochemical follow-up. J Clin Psychopharmacol 2:312–314, 1982

Weil-Malherbe H: The concentration of adrenaline in human plasma and its relation to mental activity. J Ment Sci 101:733–755, 1955

Wojcik JD, Gelenberg AJ, LaBrie RA, et al: Prevalence of tardive dyskinesia in an outpatient population. Compr Psychiatry 21:370–380, 1980

Yamaguchi T, Nagatsu T: Effects of tyrosine administration on serum biopterin in normal controls and patients with Parkinson's disease. Science 219:75–76, 1983

Young SN, Davis BA, Gauthier S: Precursors and metabolites of phenylethylamine, m- and p-tyramine and tryptamine in human lumbar and cisternal cerebrospinal fluid. J Neurol Neurosurg Psychiatry 45:633–639, 1982

Chapter 8

Decreased Tyrosine Transport in Schizophrenic Patients

Lars Bjerkenstedt, M.D.

Chapter 8

Decreased Tyrosine Transport in Schizophrenic Patients

O n the basis of the effects of neuroleptics and amphetamine on central dopaminergic mechanisms, an alteration of dopaminergic transmission in schizophrenia has been postulated (Carlsson 1978). There are also indications of disturbed dopamine metabolism in the brain of untreated schizophrenic patients (Bowers 1974). The pathophysiological mechanism for this disturbance is so far unknown.

The catecholamine dopamine is derived from the amino acid tyrosine. In the first step in this metabolic pathway, tyrosine is converted to the amino acid dihydroxyphenylalanine (L-dopa). This enzymatic reaction is catalyzed by tyrosine hydroxylase (Musacchio and Carviso 1973; Udenfriend 1966). The aromatic amino acid decarboxylase converts L-dopa to dopamine (Dairman et al. 1973). Dopamine is converted to homovanillic acid (HVA), which is the end metabolite in humans.

Central dopamine turnover is partly regulated by the availability of the precursor amino acid, tyrosine, in plasma (Fernstrom and Faller 1978; Fernstrom and Wurtman 1972).

The postabsorbtive levels of amino acids in plasma are determined by the net balance between release from endogenous protein stores, mainly muscle, and utilization by various tissues, primarily the liver; they are also influenced by the transport systems facilitating the transport of amino acids across cell membranes (Hagenfeldt et al. 1983b). Tyrosine uses a common transport system, the L-system, with other large neutral amino acids (LNAAs) (tryptophan, phenylalanine, valine, leucine, and isoleucine). The brain is the only organ for which tyrosine transport is limited at physiological plasma concentrations so that transport over the blood-brain barrier occurs in competition with other neutral amino acids (Fernstrom and Faller 1978; Pardridge and Oldendorf 1977).

163

Recently, the transport of amino acids from plasma to brain was found to be regulated by a β-adrenergic mechanism. Accordingly, the β-adrenergic agonist isoprenaline increases brain concentrations of most LNAAs, while this effect is inhibited by the β-blocking agent propranolol (Eriksson 1985).

To find out whether there is an altered availability of tyrosine in schizophrenia, two studies were conducted. In the first study (Bjerkenstedt et al. 1985), concentrations of amino acids in plasma and HVA in the cerebrospinal fluid (CSF) were determined. In the second, the transport of tyrosine in fibroblasts was studied in vitro (Hagenfeldt et al. 1987).

METHODS

With ethical approval, concentrations of the amino acids in plasma and HVA in CSF were determined in 37 drug-free schizophrenic patients (21 men and 16 women) and 65 healthy volunteers (50 men and 15 women). In the other study, 10 male schizophrenic patients participated. Five of these patients were drug-free at the time of biopsy; one of them had never received neuroleptics. Control fibroblast lines were taken from healthy female and male members of the laboratory staff and students. The patients fulfilled the research diagnostic criteria for schizophrenia. All subjects were physically healthy at routine examination, and none had a history of severe head injury or somatic disease.

The samples of blood and CSF were taken between 8 and 9 A.M. after the patients and the healthy volunteers had been at bed rest for at least 8 hours and fasting for 12 hours.

Amino acids were determined by ion-exchange chromatography (Hagenfeldt et al. 1983a) and HVA in the CSF according to the mass fragmentographic procedure described by Swahn et al. (1976). Amino acid transport in fibroblasts was measured by a cluster-tray method (Gazzola et al. 1981).

RESULTS AND COMMENTS

Amino Acids in Plasma

The schizophrenic patients had significantly higher plasma concentrations of taurine ($P < .001$), alanine ($P < .001$), valine ($P < .05$), methionine ($P < .05$), isoleucine ($P < .001$), leucine ($P < .01$), phenylalanine ($P < .01$), and lysine ($P < .001$), and lower concentrations of glutamine ($P < .05$) and histidine ($P < .05$; Table 8-1).

Table 8-1. Means and standard deviations for amino acids in plasma (μmol/l)

| Amino acids | Healthy volunteers | | | | Schizophrenic patients | | | | |
| | Men (n = 50) | | Women (n = 15) | | Men (n = 21) | | Women (n = 16) | | |
	M	SD	M	SD	M	SD	M	SD	P
Taurine	61	17	42	12	67	20	68	14	.001
Glutamine	632	83	518	89	572	116	517	78	.05
Alanine	320	87	346	64	392	111	355	113	.001
Valine	257	39	210	19	273	45	232	44	.05
Methionine	25	6	22	4	27	7	26	7	.05
Isoleucine	68	11	53	7	77	14	64	15	.001
Leucine	159	23	120	11	171	30	141	26	.01
Phenylalanine	60	7	52	5	64	8	57	8	.01
Lysine	189	27	161	21	206	29	186	28	.001
Histidine	90	9	86	9	84	11	84	14	.05

Note. Data are given only for those amino acids that differed significantly between patients and controls.

HVA in CSF

In both sexes, the HVA levels were significantly lower in the schizophrenic patients ($P < .01$). For HVA levels in CSF, a sex difference was observed, with higher concentrations in females ($P < .01$; Figure 8-1).

Relationship of Plasma Amino Acids and HVA in CSF

Plotting data on plasma amino acids against HVA in CSF showed that males and females were distributed along the same regression line for both controls and schizophrenic patients. Thus, data from both sexes were pooled in the calculations of the correlation coefficients.

Low and nonsignificant correlations were obtained between the plasma amino acid concentrations and the monoamine metabolite HVA in the control group. In contrast, the HVA level of the schizophrenic patients was significantly and negatively correlated to the branched amino acids (valine, isoleucine, and leucine) and phenylalanine and lysine (Table 8-2).

The deviations in plasma amino acids observed in the schizophrenic patients could be expected to lead to a diminished uptake of tyrosine to the brain, i.e., a reduced transport across the blood-brain barrier, and may be of pathophysiological importance for the development of the disease. This hypothesis is supported by the observations of significantly lower HVA concentrations in the schizophrenic patients

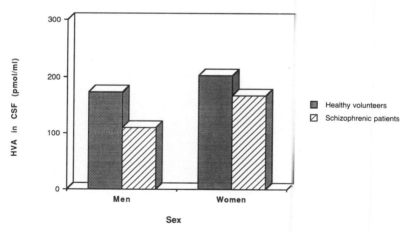

Figure 8-1.   Concentrations of HVA in CSF of schizophrenic patients and healthy volunteers.

than in the controls (Bjerkenstedt et al. 1985; Lindström 1985), and the significantly negative correlation in the patients between the concentrations of HVA and the plasma levels of those amino acids that are competing with tyrosine for transportation over the blood-brain barrier (Bjerkenstedt et al. 1985). In the controls, the correlation was low and nonsignificant.

The hypothesis of a reduced tyrosine transport in schizophrenia is supported by the finding of a significantly reduced V_{max} for tyrosine transport in vitro in fibroblasts from schizophrenic patients (Table 8-3; Hagenfeldt et al. 1987). In 9 of the 10 cell lines from the patients,

Table 8-2. Product-moment correlations between concentrations of HVA in CSF and amino acids in plasma in healthy volunteers and schizophrenic patients

Amino acids	Healthy volunteers	Schizophrenic patients
Taurine	−0.03	0.07
Aspartate	0.01	−0.21
Threonine	0.13	−0.29
Serine	0.06	−0.06
Aspargine	−0.01	−0.06
Glutamate	0.00	−0.09
Glutamine	−0.08	−0.30
Proline	0.05	−0.15
Glycine	0.16	−0.14
Alanine	0.19	−0.11
Alpha-aminobutyrate	0.00	−0.27
Valine	−0.05	−0.36*
Cystine	0.13	0.02
Methionine	0.04	−0.36*
Isoleucine	−0.06	−0.44**
Leucine	−0.15	−0.51***
Tyrosine	0.06	−0.19
Phenylalanine	−0.07	−0.35*
Ornithine	−0.18	−0.08
Lysine	−0.14	−0.33*
Histidine	0.00	0.17
Tryptophan	0.00	0.12
Arginine	0.02	−0.15

*$P < .05$. **$P < .01$. ***$P < .001$.

the V_{max} for tyrosine was below the lowest value observed among the control lines. The K_m for tyrosine in cells from the schizophrenic patients did not differ from that of control cells. For the other amino acids sharing the same transport system, as well as for glycine, which is transported independently of the others, no significant differences were observed in either of the kinetic parameters. Tyrosine transport was profoundly inhibited by the addition of one of the other amino acids sharing the L-system. There was no sodium-dependent component of the tyrosine transport in any of the cells. Thus, the isolated decrease in the capacity for tyrosine transport observed in fibroblasts from schizophrenic patients could not be explained in terms of known amino acid transport systems. It appears more likely that the present findings reflect a more general alteration in plasma membrane function in schizophrenia. Membrane abnormalities have earlier been implicated in the etiology of schizophrenia. Furthermore, changes in phospholipid composition of erythrocyte membranes (Hitzemann et al. 1984), increased permeability of muscle cell membranes (Meltzer et al. 1980), and morphological changes in skeletal muscle biopsies (Borg et al. 1987) have been demonstrated. A change in membrane characteristics may have different functional consequences depending on the cell type. Neurons, particularly dopaminergic neurons, may be more susceptible to disturbances in tyrosine transport, resulting in a decreased dopamine synthesis in these cells. This could lead to compensatory changes in dopaminergic transmission.

These findings may have implications for the study of the genetics of schizophrenia and for designing new therapeutic approaches.

Table 8-3. Mean V_{max} (nmol $\cdot$ min^{-1} $\cdot$ [mg protein]$^{-1}$) of amino acid transport in cultured fibroblasts from schizophrenic patients and controls

Amino acids	Schizophrenic patients		Healthy controls		
	M	SD	M	SD	P
L-Tyrosine	4.5	0.9	7.8	2.1	.01
L-Phenylalanine	4.8	0.6	4.2	1.1	ns
L-Tryptophan	4.6	0.9	4.6	1.1	ns
L-Leucine	8.9	1.0	8.0	1.9	ns
Glycine	4.9	2.7	3.9	1.1	ns

REFERENCES

Bjerkenstedt L, Edman G, Hagenfeldt L, et al: Plasma amino acids in relation to cerebrospinal fluid monoamine metabolites in schizophrenic patients and healthy controls. Br J Psychiatry 147:276–282, 1985

Borg J, Edström L, Bjerkenstedt L, et al: Muscle biopsy findings, conduction velocity and refractory period of single motor nerve fibres in schizophrenia. J Neurol Neurosurg Psychiatry 150:1655–1664, 1987

Bowers MB Jr: Central dopamine turnover in schizophrenic syndromes. Arch Gen Psychiatry 31:50–54, 1974

Carlsson A: Antipsychotic drugs, neurotransmitters, and schizophrenia. Am J Psychiatry 135:164–173, 1978

Dairman W, Christenson J, Udenfriend S: Characterization of dopa decarboxylase, in Frontiers in Catecholamine Research. Edited by Usdin E, Snyder S. New York, Pergamon Press, 1973, p 61

Eriksson T: Regulation of monoamine-precursor transport into the brain. Thesis, University of Gothenburgh, Gothenburgh, Sweden, 1985

Fernstrom JD, Faller DV: Neutral amino acids in the brain: changes in response to food ingestion. J Neurochem 30:1531–1538, 1978

Fernstrom JD, Wurtman RJ: Brain serotonin content: physiological regulation by plasma neutral amino acids. Science 178:414–416, 1972

Gazzola GC, Dall'Asta V, Franchi-Gazzola R, et al: The cluster-tray method for rapid measurement of solute fluxes in adherent culture cells. Anal Biochem 115:368–374, 1981

Hagenfeldt L, Bjerkenstedt L, Edman G, et al: Amino acids and monoamine metabolites in cerebrospinal fluid—interrelationships in healthy subjects. J Neurochem 42:833–837, 1983a

Hagenfeldt L, Eriksson LS, Wahren J: Amino acids in liver disease. Proc Nutr Soc 42:497–506, 1983b

Hagenfeldt L, Venizelos L, Bjerkenstedt L, et al: Decreased tyrosine transport in fibroblasts from schizophrenic patients. Life Sci 41:2749–2757, 1987

Hitzemann R, Hirschowitz J, Garver D: Membrane abnormalities in the psychoses and affective disorder. J Psychiatr Res 18:319–326, 1984

Lindström LH: Low HVA and normal 5-HIAA CSF levels in drug-free schizophrenic patients compared to healthy volunteers: correlations to symptomatology and family history. Psychiatry Res 14:265–273, 1985

Meltzer HY, Ross-Stanton J, Schlessinger S: Mean serum creatine kinase activity in patients with functional psychoses. Arch Gen Psychiatry 37:650–655, 1980

Musacchio JM, Carviso GL: Properties of tyrosine hydroxylase, in Frontiers in Catecholamine Research. Edited by Usdin E, Snyder S. New York, Pergamon Press, 1973, p 47

Pardridge WM, Oldendorf WH: Transport of metabolic substrates through the blood-brain barrier. J Neurochem 28:5–12, 1977

Swahn CG, Sandgärde B, Wiesel FA, et al: Simultaneous determination of the three major monoamine metabolites in brain tissue and body fluid by a mass fragmentographic method. Psychopharmacologia 48:147–152, 1976

Udenfriend S: Tyrosine hydroxylase. Pharmacol Rev 18:43–51, 1966

Chapter 9

Plasma and Cerebrospinal Fluid Amino Acids in Schizophrenia

C. J. A. Taylor, M.B.B.S., B.Sc., M.R.C.P.
Michael A. Reveley, M.D., Ph.D., F.R.C.P.

Chapter 9

Plasma and Cerebrospinal Fluid Amino Acids in Schizophrenia

Amino acids have been of interest in studies of the causes of schizophrenia for at least 50 years. The accumulated evidence is that disturbed amino acid metabolism might have a role in either the pathogenesis or the etiology of this common and debilitating illness. It is probable that there are several routes to developing schizophrenia, and that while one major vulnerability factor (for example, a genetic predisposition) is a necessary prerequisite, others (such as birth trauma) are also influential in determining clinical expression of the illness.

It is not surprising, therefore, that the studies of amino acids in relation to schizophrenia discussed here are wide ranging. Some examine the role of amino acids as precursors of catecholamine neurotransmitters. A substantial body of evidence exists to show that defects of dopamine metabolism, for example, are important. It has been speculated that the raised phenylalanine levels seen in phenylketonuria (PKU) predispose to psychosis, either via the dopamine system or as a more direct effect. The amino acids gamma-aminobutyric acid (GABA) and glutamate, neurotransmitters in their own right, have also been implicated. Yet other amino acids have been used to study possible abnormalities of the blood-brain barrier and hence transport of substances into the brain. The relationship of these defects, as revealed by examination of amino acids in the cerebrospinal fluid (CSF), to studies suggestive of structural brain damage is discussed. Finally, similarities in fasting amino acid levels in schizophrenia and celiac disease indicate some relationship to a medical illness with a genetic component.

173

AMINO ACID PRECURSORS OF THE CATECHOLAMINES

Hypotheses associating disordered catecholamine metabolism with schizophrenia have also implicated the precursor amino acids. There is much evidence from pharmacological and ligand binding studies in schizophrenia for excessive dopaminergic activity in the brain. The so-called dopamine hypothesis of schizophrenia was initially based on the observation that dopamine agonist drugs such as amphetamine can result in a paranoid psychosis in previously normal individuals (Connell 1958). Further, it has been demonstrated that the clinical effectiveness of antipsychotic drugs was due to blockade of postsynaptic dopamine receptors (Johnstone et al. 1978). Studies of dopamine metabolites in the CSF (Berger et al. 1980; Post et al. 1975) and in postmortem brains of patients with schizophrenia (Bird et al. 1979) have, however, failed to provide evidence of increased dopamine turnover.

Support for a role for dysfunctional noradrenergic systems in schizophrenia is less strong, relying on an unreplicated report of a reduction in dopamine-beta-hydroxylase (DBH) activity (the enzyme that synthesizes noradrenaline from dopamine) in postmortem brains (Wise et al. 1974). Woolley and Shaw (1954) hypothesized either a deficit or an excess of serotonergic function in schizophrenia after investigating potential hypotensive drugs that were active on 5-hydroxytryptamine (5-HT) systems. Some of these drugs not only had some structural resemblance to 5-HT, but had behavioral effects in animals and caused mental disturbances in humans. There has been little evidence for abnormal 5-HT turnover from studies of CSF metabolites (Ashcroft et al. 1966; Berger et al. 1980; Persson and Roos 1969), from studies of postmortem schizophrenic brains (Joseph et al. 1979), or from 5-HT receptor ligand binding studies (Owen et al. 1981; Whitaker et al. 1981).

Dopamine and noradrenaline are formed in the brain from their amino acid precursor tyrosine; tyrosine is converted by tyrosine hydroxylase to L-dopa, which is in turn converted to dopamine by dopa decarboxylase. Tyrosine, however, is formed by the action of phenylalanine hydroxylase on phenylalanine in the liver, and it has been suggested that an excess of phenylalanine peripherally could lead to a corresponding excess of catecholamines centrally. Tryptophan, a precursor of serotonin, has been little studied in relation to schizophrenia. Administration of 5-hydroxytryptophan, the immediate precursor of 5-HT, in conjunction with a peripheral decarboxylase inhibitor (the combined effect being to raise brain levels of 5-HT) had either variable or no effects on symptoms (Wyatt et al. 1972).

Phenylalanine was first associated with schizophrenia by Penrose

(1935), who reported six cases of depressive psychosis in the family of a patient with PKU. He suggested that the heterozygous state for PKU might predispose to schizophrenia. This is the most common autosomal recessive disorder, and it results in a lack of liver phenylalanine hydroxylase (PAH), such that ingested phenylalanine cannot be metabolized. Untreated, this condition leads to mental handicap in homozygous individuals, and some have noted a high prevalence of psychotic illness in addition (Pitt 1971). Heterozygotes (who have reduced PAH activity and hence somewhat raised peripheral phenylalanine levels) have also been reported as suffering from a higher rate of psychotic illness. A study of 40 schizophrenic inpatients by Poisner (1960) found that the mean fasting serum phenylalanine was significantly higher in patients than in controls, suggesting a higher prevalence of heterozygosity for PKU in the patient population. Perry et al. (1973b) found one case of un-suspected PKU by screening inmates of a large mental hospital, but a more recent study of schizophrenic inpatients failed to find any associated occult PKU (Reveley and Reveley 1982).

Some family studies have lent support to Penrose's hypothesis (e.g., Thompson 1957) by finding an increased frequency of psychosis in PKU families. Perry et al. (1973b) reported two cases of adult PKU in siblings presenting as psychotic illness: in a sibship of four adult sufferers from previously unrecognized and untreated PKU, three were of normal intelligence, but two had suffered major psychotic illnesses resulting in their admission to a psychiatric hospital. Other family studies (Larson and Nyman 1968; Perry 1966) did not confirm these observations in PKU families, but noted the predominance of affective symptoms in those relatives who did suffer from schizophrenia. The studies postulated that heterozygosity for the PKU gene is a modifying factor in the clinical presentation of psychosis. Kuznetsova (1972) studied the frequency and phenotypic manifestation of schizophrenia in the parents of PKU patients, i.e., known heterozygotes. He identified 150 affected families, and among them found an age-corrected incidence of 2.4% of schizophrenia in the 300 parents, as compared to a population incidence of 0.85% in Europe. He notes, however, that these parents are unlikely to be representative of the general population, since they had married and had children. The severest forms of schizophrenia, with early onset and reduced fertility, must have been excluded. The clinical picture was generally one of a benign, slowly progressive disease with a high frequency of affective symptoms of a depressive nature. He concludes that this phenotype could be due to the modifying presence of the PKU gene on the schizophrenia gene.

Poisner (1960) also postulated a role for disordered phenylalanine metabolism in schizophrenia, linking this to evidence presented by Kallman as long ago as 1938 for a genetic component to the illness. It is of interest that the gene for PAH, located on chromosome 12, has been suggested as a "candidate gene" for molecular genetic studies of psychotic illnesses (Gurling 1986).

THE ROLE OF NEUROTRANSMITTER AMINO ACIDS

Gamma-Aminobutyric Acid

Gamma-aminobutyric acid is well established as an inhibitory neurotransmitter controlling the opening of chloride ion channels (Tallman et al. 1980). It is formed from its precursor amino acid L-glutamate by the action of glutamic acid decarboxylase (GAD), an enzyme specifically present in GABA nerve endings. These nerve endings constitute about 30% of all brain synapses, making GABA a relatively widespread substance. GABA is also related to carbohydrate metabolism: glucose is another precursor, and one of its degradation products, succinic semialdehyde, is metabolized to succinic acid, which is in turn involved in Kreb's cycle. GABA receptors are present in highest density in the hippocampus, cerebral cortex, and cerebellum (Enna et al. 1977). Two kinds of GABA receptor have been demonstrated: "high-affinity" sites bind to tritiated muscimol and tritiated GABA, and "low-affinity" sites appear to be under the modulatory control of a benzodiazepine binding site (Tallman et al. 1980). GAD is present in high concentrations in parts of the basal ganglia and the substantia nigra (Vogel et al. 1975).

Interest in the possible role of GABA in the pathogenesis of schizophrenia developed when a relationship between GABA-ergic and dopaminergic neurons was demonstrated. Thus there is much evidence for long GABA feedback loops in the nigrostriatal dopamine pathway (Kim et al. 1971; Lloyd 1978), and in the mesocortical and mesolimbic dopamine pathways (Fuxe et al. 1977). Stevens et al. (1974) induced "paranoid, psychosis-like" behavior in rats by infusing a GABA antagonist into the area of the nucleus accumbens. The suggestion was that GABA may act presynaptically and/or postsynaptically (via interneurons regulating dopamine transmission) to inhibit dopamine release. Dopamine-blocking agents such as haloperidol have been shown to decrease the turnover time of GABA in the substantia nigra, caudate, and nucleus accumbens (Marco et al. 1976).

It has been hypothesized that, in schizophrenia, the GABA system may be defective (Roberts 1976), leading to the excessive dopaminergic activity postulated by the dopamine theory of schizophrenia. Another possibility is that there is an interaction between alterations in GABA activity and dopamine receptor sensitivity (van Kammen 1977). Dopamine receptor hypersensitivity has also been implicated in schizophrenia (Owen et al. 1978). The finding by Bird et al. (1977) of significant reductions in GAD activity in schizophrenic brains seemed at first to lend support to Roberts' hypothesis, but it was not confirmed by two later reports (Crow et al. 1978; Perry et al. 1978). It was then discovered that the low GAD activity was in fact the result of an excess of patients in the schizophrenic group dying of bronchopneumonia, the terminal hypoxia having the effect of reducing GAD activity (Bird et al. 1978).

There have been numerous studies of CSF GABA levels in schizophrenia, although reports of low levels in other neuropsychiatric disorders such as Huntington's chorea and Alzheimer's disease (Enna et al. 1980) reveal a nonspecific (as yet unestablished) relationship to biochemical and structural abnormality. Levels in schizophrenic patients have been reported as low (van Kammen et al. 1981), high (McCarthy et al. 1981), and normal (Gerner and Hare 1981). The relationship of CSF GABA levels to clinical variables such as age, premorbid functioning, and neuroleptic treatment is as yet unknown.

A more recent study of 30 drug-free patients showed that women with schizophrenia had significantly lower GABA levels than age- and sex-matched normal controls. Levels increased with duration of illness, number and duration of hospitalizations, and age (van Kammen et al. 1982). The authors suggest that low GABA levels in schizophrenia are particularly characteristic of young, recently ill female patients. Levels in the latter were similar to those of patients in another study by Gerner and Hare (1981). Van Kammen et al. (1982) also attempted to relate CSF GABA levels to clinical features of the illness. A nonsignificant correlation with psychosis ratings was found, as well as a relationship between GABA and blunted affect. Bowers et al. (1980) also report low GABA levels to be associated with negative symptomatology (emotional withdrawal). Table 9-1 compares the results of the various studies.

Postmortem studies may be of more value in showing localized deficits, although methodological problems connected with GAD values and hence GABA levels after death are difficult to overcome. Brain GAD levels and GABA receptor binding in the cortex have been found by some authors to be normal in schizophrenia (Bennet et al. 1979), as have GABA levels themselves (Cross et al. 1979). Others, however, have found GABA levels to be decreased. Perry et al. (1979a) compared levels in postmortem brains of patients with schizophrenia and Huntington's chorea to those of normal controls. They found reductions of mean GABA content in the nucleus accumbens and thalamus in both patient groups, although they point out the wide variation in individual GABA levels and the fact that reductions were not seen in all patients. This finding did not seem to be related to treatment with antipsychotic drugs, and supports Roberts' theory of defective modulation of dopaminergic neurons by GABA systems leading to dopaminergic overactivity in schizophrenia. We note that the clinical efficacy of butyrophenones used in the treatment of schizophrenia correlates positively with their relative potencies in inhibiting presynaptic GABA uptake in rat brain. The development of techniques such as positron-emission tomography (PET) will make possible in vivo study of receptor sites throughout the progression of the illness and its treatment without these methodological problems. Radiolabeled ligands such as [^{11}C]raclopride (a highly selective D2 dopamine antagonist) have been used in studies of schizophrenic patients to shed light on abnormalities of dopamine receptor density (Farde et al. 1987). The application of similar techniques to GABA receptor sites will clarify the role of GABA, if any, and the nature of any interaction with dopamine systems in the illness.

Glutamic Acid

Glutamic acid, present in high concentrations in the brain, has also been implicated in schizophrenia. It has been shown to be a transmitter of corticostriate neurons (Kim et al. 1977) and of hippocamposeptal projections (Zaczek et al. 1979). As the precursor of GABA,

Table 9-1.   Studies of GABA and GAD levels in schizophrenia

Authors	Tissue	Amino acid/enzyme	Finding
Bird et al. 1977	PmB	GAD	Reduced
Crow et al. 1978	PmB	GAD	Normal
Perry et al. 1978	PmB	GAD	Normal
Bennet et al. 1979	PmB	GAD	Normal
Cross et al. 1979	PmB	GABA	Normal
Perry et al. 1979a	PmB	GABA	Reduced
Gerner and Hare 1981	CSF	GABA	Normal
McCarthy et al. 1981	CSF	GABA	Raised
van Kammen et al. 1981	CSF	GABA	Reduced
van Kammen et al. 1982	CSF	GABA	Reduced

Note. PmB = postmortem brain.

glutamate may also have a modulatory role on the dopaminergic systems thought to be disordered in the illness. It is of interest that patients with Huntington's chorea, which is often associated with a schizophrenia-like psychosis in the early stages, have been reported to have significantly reduced levels of CSF glutamic acid (Kim et al. 1980a). We found a similar reduction in the CSF of schizophrenic patients, although plasma levels of glutamate in both illnesses were similar to those of normal controls. Table 9-2 compares results of studies of glutamate levels.

A hypothesis arose from these findings that glutamatergic neurons in schizophrenia had degenerated or were otherwise functionally impaired, leading to decreased glutamate release as reflected by low CSF levels. Animal studies seem to support this suggestion: chronic amphetamine administration to rats (intended to simulate the dopaminergic hyperfunction and psychotic state this is known to induce in humans) resulted in lower CSF glutamate and increased brain glutamate concentrations in several areas of rat brain (Kim et al. 1981). The hypothesis was not supported, however, by Perry (1982), who found no differences in glutamate levels in CSF and postmortem brains of schizophrenic patients when compared with controls. Perry et al. (1979b) have also reported that daily injection of rats with large doses of chlorpromazine or haloperidol over a period of 100 days produced no alteration of glutamic acid content in the mesolimbic area of the brain. Thus the extent to which a possible hypofunction of glutamatergic neurons interacts with a disorder of dopaminergic systems in schizophrenia remains unclear.

ABNORMAL AMINO ACID METABOLISM

Pepplinkhuizen et al. (1980) report instances of psychoses being induced in susceptible individuals given oral loading doses of serine, although abnormalities of the blood-brain barrier were not studied at the same time. They investigated four patients with an intermittent psychotic illness closely resembling that induced by the hallucinogenic methylated indolamines such as those related to lysergic acid diethylamide (LSD). The transmethylation hypothesis of schizophrenia

Table 9-2.   Glutamate levels in schizophrenia

Authors	Tissue	Finding
Kim et al. 1980a	CSF	Reduced
"	Plasma	Normal
Perry 1982	CSF	Normal
"	Postmortem brain	Normal

suggested that some of the symptoms might be due to the abnormal accumulation of a methylated biogenic amine with psychotogenic properties similar to those of LSD. The patients were suspected of having a porphyric illness, and all patients were noted to have decreased serine excretion during the psychotic episodes.

We suggest an accelerated conversion of serine into glycine (because the increase in pyrrol synthesis seen in porphyria demands increasing amounts of glycine). The excess of methyl groups produced by this conversion could result in faulty methylation of catecholamines (Osmond and Smythies 1952). All patients reported by Pepplinkhuizen et al. (1980) reacted to an oral loading dose of serine with a short-lived psychosis of rapid onset. One patient reacted to glycine in the same way. The suggestion was that disturbed serine and glycine metabolisms have a central role in the development of schizophrenia-like symptoms, not just via the production of abnormally methylated monoamines, but possibly by some more direct toxic effect.

ABNORMAL AMINO ACID TRANSPORT MECHANISMS

Another aspect of amino acid metabolism studied in schizophrenia is their transport into the brain. Neutral amino acids, including the monoamine precursors tyrosine and tryptophan, compete with each other for a common transport system, the L-system. Since central monoamine turnover is partly influenced by the availability of these precursor amino acids in plasma (Fernstrom and Faller 1978), factors affecting this transport system may have an indirect effect on central monoamine turnover.

One study of plasma levels of amino acid in schizophrenic patients was by Bjerkenstedt et al. (1985). Compared to normal controls, the patients had higher plasma levels of taurine, methionine, valine, isoleucine, leucine, phenylalanine, and lysine. Except for taurine, these all share the L-transport system for neutral amino acids into the brain. Plasma levels of tyrosine and tryptophan (the precursors of dopamine and serotonin) were unchanged. In addition, it was also found that CSF levels of the dopamine metabolite homovanillic acid (HVA) and the serotonin metabolite 5-hydroxyindoleacetic acid (5-HIAA) were lower in patients than in controls. There was in fact a negative correlation between plasma amino acid levels and CSF HVA and 5-HIAA levels.

We postulate that raised levels of plasma amino acids compete with tyrosine and tryptophan for entry into the brain via the L-transport system. This would lead to a reduced brain uptake of these two amino acids, and hence a lower production of the relevant monoamine transmitters. This lower production is reflected in the finding of reduced CSF dopamine and serotonin metabolites. A compensatory increase in dopamine receptor sensitivity would make these findings compatible with the hypothesis of increased dopamine transmission in schizophrenia.

Other studies of CSF monoamine metabolites have failed to show evidence for increased turnover of dopamine or serotonin, since CSF levels of HVA and 5-HIAA have been normal (Berger et al. 1980; Persson and Roos 1969; Post et al. 1975). Bjerkenstedt et al. argue that this is possibly due to methodological differences, disease heterogeneity, and lack of adequate control groups. They also report, in the healthy controls, a positive correlation between CSF hydroxy-3-methoxyphenylethylene glycol (HMPG) and several plasma amino acids. This relationship was disturbed in the schizophrenic group for unknown reasons.

EVIDENCE OF IMPAIRMENT OF THE BLOOD-BRAIN BARRIER

It has been suggested that an impairment in the blood-brain barrier may be associated with certain psychiatric illnesses. Axelsson et al. (1982) investigated the prevalence and clinical significance of such a defect in patients with paranoid psychosis. They established impairment in the blood-brain barrier by detecting an increased ratio between the albumin concentrations in CSF and serum, allowing for the fact that this ratio normally increases with age. An increase in this ratio was detected when comparing patients with paranoid psychosis to normal controls. A highly significant finding was that the onset of psychosis had occurred, on average, 20 years earlier in the patients with such an impairment in the blood-brain barrier than in those without. The effect did not seem to be related to drug treatment. We suggest that impairment of the blood-brain barrier is causally related to the illness, possibly by allowing entry of substances toxic to the brain and leading to psychosis in susceptible individuals. Another possibility is that of an interrelationship with the disturbed transport of amino acids into the brain.

RELATIONSHIP OF AMINO ACIDS TO STRUCTURAL ABNORMALITY IN THE BRAIN

Very few studies have attempted to relate any abnormalities of amino acids to brain structural defects found in schizophrenia, although there have been some reports associating such defects with dopamine metabolites. A fairly consistent finding in schizophrenia is that of enlarged cerebral ventricles (e.g., Johnstone et al. 1976), and some studies have found that this abnormality correlates negatively with levels of dopamine metabolites such as HVA in the CSF (van Kammen et al. 1983; Nybäck et al. 1983). These findings of lower monoamine metabolite levels in patients with ventricular enlargement suggest that enlargement may be due to cerebral atrophy involving damage to neurons involved in monoamine pathways.

One study (Reveley et al. 1987) examined the relationship of CSF amino acid concentrations to ventricular enlargement. Significantly higher CSF levels of alanine, glycine, leucine, and phenylalanine were found in schizophrenic patients as compared to controls. These results confirmed the finding by Bjerkenstedt et al. (1985) of raised CSF leucine levels in patients with schizophrenia, but not the finding of raised CSF histidine. Another similarity in the findings of the two studies was that the amino acids involved were all neutral. Of those patients in the schizophrenic group (Reveley et al. 1987) found to have ventricular enlargement, a high correlation was found with elevated CSF alanine. We hypothesize that brain damage, as reflected by dilatation of the cerebral ventricles, may cause defects in the amino acid transport mechanisms or to the blood-brain barrier, leading to the observed rise in CSF amino acids.

Reveley et al. (1987) also note that studies of other diseases associated with degeneration of central neurons have found evidence of raised CSF levels of a variety of amino acids. High levels of alanine, glycine, phenylalanine (as in the schizophrenic patients), and of methionine, valine, and threonine (unlike the schizophrenic patients) have been found in amyotrophic lateral sclerosis (ALS) (de Belleroche et al. 1984). Table 9-3 compares the findings of several studies.

In Parkinson's disease and other extrapyramidal disorders, Lakke and Teelken (1976) also found significantly elevated alanine, glycine, leucine, and phenylalanine, and in addition, elevated valine, isoleucine, tyrosine, citrulline, and histidine. In the study of ALS patients, alanine levels were found to increase with the duration of the disease, while glycine levels correlated significantly with the activity of the disease (as measured by severity/duration). Reveley and Reveley (1982) cite evidence for ventricular enlargement being progressive in at least some schizophrenic patients (Nasrallah et al. 1986) and suggest that rescan follow-up studies might be valuable in conjunction with investigation of amino acid and other metabolic defects in elucidating the extent to which schizophrenia is a progressive deterioration.

AMINO ACID ABNORMALITIES IN SCHIZOPHRENIA AND CELIAC DISEASE

Several authors have noted that the psychiatric symptoms sometimes seen in celiac disease are similar to those of schizophrenia. Dohan (1966) postulated a partial genetic relationship between the two illnesses, because some epidemiological data support an association between them (Baldwin 1979). This association has not always been confirmed, however (Jablensky 1986). There have been many conflicting reports as to the beneficial effects of a gluten-free diet on psychotic symptoms (Rice et al. 1978; Storms et al. 1982) and on the exacerbating effects of a gluten challenge (Potkin et al. 1981; Singh and Kay 1976).

Manowitz (1978) suggested that if schizophrenia and celiac disease share common etiological factors, this might be reflected in biochemical measures such as fasting amino acid plasma levels. He presents evidence from a study of levels in schizophrenic patients (Perry et al. 1973a) compared with levels in a study of celiac patients (Douglas and Booth 1969) that similarities do exist. In order to compare results in the two studies, he calculated the difference between patient (P) and control (C) mean amino acid levels, and divided this difference by the control value (P–C)/C for each amino acid. He found the P–C/C values of patients with schizophrenia to be positively correlated with those of both treated and untreated celiac patients. He argues that this lends support to the hypothesis that the illnesses share

Table 9-3. Raised CSF levels of amino acids in schizophrenia and other neurodegenerative diseases

Authors	Amino acids	Disease
Lakke and Teelken 1976	Alanine, glycine, leucine, isoleucine, phenylalanine, valine, tyrosine, citrulline, histidine	Parkinson's disease
de Belleroche et al. 1984	Alanine, glycine, phenylalanine, methionine, valine, threonine	ALS
Bjerkenstedt et al. 1985	Leucine, isoleucine, histidine	Schizophrenia
Reveley et al. 1987	Alanine, glycine, leucine, phenylalanine	Schizophrenia

a common predisposing factor, rather than resulting from the disease process itself. He suggests that, if schizophrenia is in fact a group of illnesses with several biochemical causes (Pfeiffer 1976), the similarities of amino acid profiles compared to those in celiac disease might be even more significant in a subset of patients with schizophrenia. As yet, we have no way of selecting such a subset, but the study of amino acid profiles and metabolism in the different diagnostic categories of schizophrenia may prove a fruitful area for future research. It is of interest that celiac disease shows a clearcut association with the HLA-B8 antigen (present in 66 to 88% of cases). This antigen has been reported in at least one study to show a significant association with the hebephrenic subtype of schizophrenia (McGuffin et al. 1981).

SUMMARY AND CONCLUSIONS

It is clear from the preceding discussion that amino acids have been implicated in schizophrenia in a number of ways. A central theme of most of the studies considered is how evidence for disordered amino acid metabolism complements and supports that for the dopamine theory of schizophrenia. Thus, raised levels of the precursor amino acid phenylalanine may lead to dopaminergic overactivity, and impaired transport of the precursor tyrosine into the brain may result in dopamine receptor hyperactivity. Faults in GABA-ergic and glutamatergic systems, thought to regulate dopaminergic systems, could play a role by either of these mechanisms.

Other evidence from studies of amino acids suggests impairment of the blood-brain barrier in schizophrenia and contributes to other findings implying a structural brain abnormality in the illness. Finally, some studies link schizophrenic symptoms to physical illnesses such as phenylketonuria and celiac disease, possibly via a genetic route.

The sheer variety of the evidence gleaned from amino acid research suggests that amino acids probably do not play a central role in the etiology of schizophrenia, but rather reflect the pathogenic process at various points along its pathway. It is likely that only when the primary etiological factor(s) of this disease is discovered will this pathway, and hence the diversity of amino acid abnormalities described here, be unraveled.

REFERENCES

Ashcroft GW, Crawford TBB, Eccleston D, et al: 5-Hydroxyindole compounds in the cerebrospinal fluid of patients with psychiatric or neurological diseases. Lancet 2:1049–1052, 1966

Axelsson R, Martensson E, Alling C: Impairment of the blood-brain barrier as an aetiological factor in paranoid psychosis. Br J Psychiatry 141:273–281, 1982

Baldwin JA: Schizophrenia and physical disease. Psychol Med 9:611, 1979

Bennet JP, Enna SJ, Bylund DB, et al: Neurotransmitter receptors in frontal cortex of schizophrenics. Arch Gen Psychiatry 36:927–934, 1979

Berger PA, Faull KF, Kilowski J, et al: Cerebrospinal fluid monoamine metabolites in depression and schizophrenia. Am J Psychiatry 137:174–180, 1980

Bird ED, Spokes EG, Barnes J, et al: Increased brain dopamine and reduced glutamic acid decarboxylase and choline acetyltransferase activity in schizophrenia and related psychoses. Lancet 2:1157–1159, 1977

Bird ED, Spokes EG, Barnes J, et al: Biochemistry in schizophrenia. Lancet 1:156, 1978

Bird ED, Crow TJ, Iverson LL, et al: Dopamine and homovanillic acid concentrations in post-mortem brain in schizophrenia. J Physiol 293:36–37P, 1979

Bjerkenstedt L, Edman G, Hagenfeldt L, et al: Plasma amino acids in relation to cerebral spinal fluid monoamine metabolites in schizophrenic patients and healthy controls. Br J Psychiatry 147:276–282, 1985

Bowers MB, Heninger GR, Sternberg D, et al: Clinical processes and central dopaminergic activity in psychotic disorders. Community Psychopharmacology 4:177–188, 1980

Connell PH: Amphetamine psychosis. Maudsley Monograph No 5. London, Chapman & Hall, 1958

Cross AJ, Crow TJ, Owen F: Gamma-aminobutyric acid in the brain of schizophrenic patients. Lancet 1:560–561, 1979

Crow TJ, Owen F, Cross AJ, et al: Brain biochemistry in schizophrenia. Lancet 1:36–37, 1978

de Belleroche J, Recordati A, Clifford Rose F: Elevated levels of amino acids in the CSF of motor neurone disease patients. Neurochem Pathol 2:1–6, 1984

Dohan FC: Cereals and schizophrenia: data and hypothesis. Acta Psychiatr Scand 42:125–152, 1966

Douglas AP, Booth CC: Post-prandial plasma free amino acids in adult coeliac disease after oral gluten and albumin. Clin Sci 37:643–653, 1969

Enna SJ, Bennett JP, Bylun DP, et al: Neurotransmitter receptor binding: regional distribution in human brain. J Neurochem 28:233–236, 1977

Enna SJ, Ziegler MG, Lake CR, et al: CSF GABA: correlation with CSF and blood constituents and alteration in neurological disorders, in Neurobiology of Cerebrospinal Fluid. Edited by Wood JH. New York, Plenum Press, 1980

Farde L, Wiesel FA, Hall H, et al: No D_2 receptor increase in PET study of schizophrenia. Arch Gen Psychiatry 44:671, 1987

Fernstrom JD, Faller DV: Neutral amino acids in the brain: changes in response to food ingestion. J Neurochem 30:1531–1538, 1978

Fuxe K, Perez de la Mora M, Hokfelt T, et al: GABA-DA interactions and their possible relation to schizophrenia, in Psychopathology and Brain Dysfunction, Vol 1. Edited by Shagass C, Gershon S, Friedhof AJ. New York, Raven Press, 1977, pp 99–111

Gerner RH, Hare TA: CSF GABA in normals, depression, schizophrenia, mania and anorexia nervosa. Am J Psychiatry 138:1098–1101, 1981

Gurling H: Candidate genes and favoured loci: strategies for molecular genetic research into schizophrenia, manic depression, autism, alcoholism and Alzheimer's disease. Psychiatr Dev 4:289–309, 1986

Jablensky A: Epidemiology of schizophrenia: a European perspective. Schizophr Bull 12:52, 1986

Johnstone EC, Crow TJ, Frith CD, et al: Cerebral ventricular size and cognitive impairment in schizophrenia. Lancet 2:924–926, 1976

Johnstone EC, Crow TJ, Frith CD, et al: Mechanism of the antipsychotic effect in the treatment of acute schizophrenia. Lancet 1:848–851, 1978

Joseph MH, Baker HF, Crow TJ, et al: Brain tryptophan metabolism in schizophrenia: a post-mortem study of metabolites on the serotonin and kynurenine pathways in schizophrenic and control subjects. Psychopharmacology (Berlin) 62:279–285, 1979

Kallman F: The Genetics of Schizophrenia. New York, Augustin, 1938

Kim JS, Bak JK, Hassler R, et al: Role of gamma-aminobutyric acid (GABA) in the extrapyramidal motor system: II. Some evidence for the existence of a type of GABA-rich strio-nigral neuron. Exp Brain Res 14:95–104, 1971

Kim JS, Hassler R, Haug P, et al: Effect of frontal ablations on striatal glutamic acid level in rat. Brain Res 132:370–374, 1977

Kim JS, Kornhuber HH, Holzmuller B, et al: Reduction of cerebrospinal fluid glutamic acid in Huntington's chorea and in schizophrenic patients. Arch Psychiatr Nervenkr 228:7–10, 1980a

Kim JS, Kornhuber HH, Schmid-Burgk W, et al: Low cerebrospinal fluid glutamate in schizophrenia and a new hypothesis of schizophrenia. Neurosci Lett 20:379–382, 1980b

Kim JS, Kornhuber HH, Brand U, et al: Effects of chronic amphetamine treatment on the glutamate concentration in cerebrospinal fluid and brain: implications for a theory of schizophrenia. Neurosci Lett 24:93–96, 1981

Kuznetsova LI: Frequency and phenotypic manifestations of schizophrenia in the parents of patients with phenylketonuria. Genetika 8(4):172–173, 1972

Lakke JPWF, Teelken AW: Amino acid abnormalities in cerebrospinal fluid of patients with Parkinsonism and extrapyramidal disorders. Neurology 26:489–493, 1976

Larson CL, Nyman GE: Phenylketonuria. Mental illness in heterozygotes. Psychiat Klin (Basel) 1:367, 1968

Lloyd KG: Neurotransmitter interactions related to central dopamine neurons, in Essays in Neurochemistry and Neuropharmacology, Vol 3. Chichester, UK, John Wiley, 1978, p 129

Manowitz P: Amino acid levels in schizophrenia: a clue to aetiology. Biol Psychiatry 13:489–491, 1978

Marco E, Mao CC, Cheney DL, et al: The effects of antipsychotics on the turnover rate of GABA and acetylcholine in rat brain nuclei. Nature 264:363, 1976

McCarthy BW, Gomes UR, Neethling AC, et al: Gamma-aminobutyric acid concentration in cerebrospinal fluid in schizophrenia. J Neurochem 36:1406–1408, 1981

McGuffin P, Farmer A, Yonace A: HLA antigens and subtypes of schizophrenia. Psychiatry Res 5:115–122, 1981

Nasrallah HA, Olson SC, McCalley-Whitters M, et al: Cerebral ventricular enlargement in schizophrenia: a preliminary follow-up study. Arch Gen Psychiatry 43:157–159, 1986

Nybäck H, Berggren B-M, Hindmarsh T, et al: Cerebroventricular size and cerebrospinal fluid monoamine metabolites in schizophrenic patients and healthy volunteers. Psychiatry Res 9:301–308, 1983

Osmond H, Smythies JR: Schizophrenia: a new approach. J Ment Sci 98:309–315, 1952

Owen F, Cross AJ, Crow TJ, et al: Increased dopamine receptor sensitivity in schizophrenia. Lancet 2:223–225, 1978

Owen F, Cross AJ, Crow TJ, et al: Neurotransmitter receptors in brain in schizophrenia. Acta Psychiatr Scand 63 (suppl 291):20–26, 1981

Penrose LS: Inheritance of phenylpyruvic amentia (phenylketonuria). Lancet 2:192, 1935

Pepplinkhuizen L, Bruinvels J, Blom W, et al: Schizophrenia-like psychosis caused by a metabolic disorder. Lancet 1:454–455, 1980

Perry TL: The incidence of mental illness in the relatives of individuals suffering from phenylketonuria or mongolism. Psychiatry Res 4:51, 1966

Perry TL: Normal cerebrospinal fluid and brain glutamate levels in schizophrenia do not support the hypothesis of glutamatergic neuronal dysfunction. Neurosci Lett 28:81–85, 1982

Perry TL, Hansen S, Lesk D, et al: Amino acids in plasma, cerebrospinal fluid and brain of patients with Huntington's chorea, in Advances in Neurology, Vol 1. Edited by Barbeau A, Chase TN, Paulson GW. New York, Raven Press, 1973a, pp 609–618

Perry TL, Hansen S, Tischler B, et al: Unrecognized adult phenylketonuria. New Engl J Med 289:395–398, 1973b

Perry TL, Blessed G, Perry RH, et al: Brain biochemistry in schizophrenia. Lancet 1:35–36, 1978

Perry TL, Buchanan J, Kish SJ, et al: Gamma-aminobutyric acid deficiency in brain of schizophrenic patients. Lancet 1:237–239, 1979a

Perry TL, Hansen S, Kish SJ: Effects of chronic administration of anti-psychotic drugs on GABA and other amino acids in the mesolimbic area of rat brain. Life Sci 24:283–288, 1979b

Persson T, Roos BE: Acid metabolites from monoamines in cerebrospinal fluid in chronic schizophrenics. Br J Psychiatry 115:95–98, 1969

Pfeiffer CC: The schizophrenias '76. Biol Psychiatry 2:773–775, 1976

Pitt D: The natural history of untreated phenylketonuria. Med J Aust 1:378–383, 1971

Poisner AM: Serum phenylalanine in schizophrenia: biochemical genetic aspects. J Nerv Ment Dis 131:74–76, 1960

Post RM, Fink E, Carpenter WT, et al: Cerebrospinal fluid amine metabolites in acute schizophrenia. Arch Gen Psychiatry 32:1063–1069, 1975

Potkin SG, Weinberger D, Kleinman J, et al: Wheat challenge in schizophrenic patients. Am J Psychiatry 138:1208–1211, 1981

Reveley AM, Reveley MA: Screening for adult phenylketonuria in psychiatric inpatients. Biol Psychiatry 17:1343–1345, 1982

Reveley MA, de Belleroche J, Recordati A, et al: Increased CSF amino acids and ventricular enlargement in schizophrenia: a preliminary study. Biol Psychiatry 2:413–420, 1987

Rice JR, Ham CH, Gore WE: Another look at gluten in schizophrenia. Am J Psychiatry 135:1417–1418, 1978

Roberts E: Disinhibition as an organizing principle in the nervous system— the role of the GABA system: application to neurologic and psychiatric disorders, in GABA in Nervous System Function. Edited by Roberts E, Chase TN, Tower TB. New York, Raven Press, 1976, pp 515–539

Singh MM, Kay SR: Wheat gluten as a pathogenic factor in schizophrenia. Science 191:401–402, 1976

Stevens JR, Wilson K, Foote W: GABA blockade, dopamine and schizophrenia: experimental studies in the cat. Psychopharmacologia 39:105–119, 1974

Storms LH, Jamie M, Clopton MS, et al: Effects of gluten on schizophrenics. Arch Gen Psychiatry 39:323–327, 1982

Tallman JF, Paul SM, Skolnick P, et al: Receptors for the age of anxiety: pharmacology of the benzodiazepines. Science 207:274–281, 1980

Thompson J: Relatives of phenylketonuric patients. J Ment Defic Res 1:67–78, 1957

van Kammen DP: Gamma-aminobutyric acid (GABA) and the dopamine hypothesis of schizophrenia. Am J Psychiatry 134:138–143, 1977

van Kammen DP, Sternberg DE, Hare TA, et al: Spinal fluid GABA levels in schizophrenia, in Biochemistry of the Human Cerebrospinal Fluid, Vol 31. Edited by Angrist B, Burrows GD, Lader M. Advances in the Bio-sciences. New York, Pergamon Press, 1981, pp 315–321

van Kammen DP, Sternberg DE, Hare TA, et al: CSF levels of gamma-aminobutyric acid in schizophrenia. Low values in recently ill patients. Arch Gen Psychiatry 39:91–97, 1982

van Kammen DP, Mann LS, Sternberg DE, et al: Dopamine-beta-hydroxylase activity and homovanillic acid in spinal fluid of schizophrenics with brain atrophy. Science 220:974–977, 1983

Vogel WH, Heginbothom SD, Boehme DH: Glutamic acid decarboxylase, glutamine synthetase and glutamic acid dehydrogenase in various areas of human brain. Brain Res 88:131–135, 1975

Whitaker PM, Crow TJ, Ferrer IN: 3-H-LSD binding in front⟍ schizophrenia. Arch Gen Psychiatry 38:278–280, 1981

Wise CD, Baden MM, Stein L: Post-mortem measuremen⟍ huma⟍ brain: evidence of a central noradre⟍ schizopt⟍renia. J Psychiatr Res 11:185–198, 1974

Woolley DW, Shaw EA: A biochemical and pharm⟍ about certai⟍ mental disorders. Proc Natl Aca⟍ 1954

Wyatt RJ, Vaughan T, Galanter M, et al: Be⟍ schizophrenic p⟍tients given L-5-⟍ 177:1124–1126, 1972

Zaczek R, Hedreen JC, Coyle JP: E⟍ glutamatergic pathway in ⟍he rat. ⟍